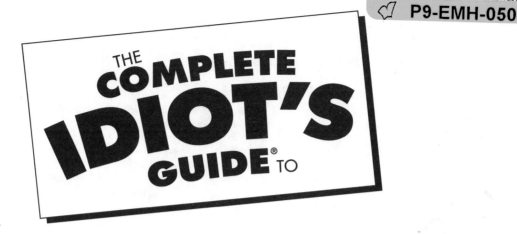

THE
COMPLETE
IDIOT'S
GUIDE® TO

Physics

by Johnnie T. Dennis

ALPHA

A member of Penguin Group (USA) Inc.

To my wife, Shirley, who walked through every page with me from first to last, and to my children and grandchildren, who supported me with prayers and encouraging words.

International Standard Book Number: 1-59257081-X
Library of Congress Catalog Card Number: 2003105466

05 04 03   8 7 6 5 4 3 2 1

Interpretation of the printing code: The rightmost number of the first series of numbers is the year of the book's printing; the rightmost number of the second series of numbers is the number of the book's printing. For example, a printing code of 03-1 shows that the first printing occurred in 2003.

Printed in the United States of America

Note: This publication contains the opinions and ideas of its author. It is intended to provide helpful and informative material on the subject matter covered. It is sold with the understanding that the author and publisher are not engaged in rendering professional services in the book. If the reader requires personal assistance or advice, a competent professional should be consulted.

The author and publisher specifically disclaim any responsibility for any liability, loss, or risk, personal or otherwise, which is incurred as a consequence, directly or indirectly, of the use and application of any of the contents of this book.

Most Alpha books are available at special quantity discounts for bulk purchases for sales promotions, premiums, fund-raising, or educational use. Special books, or book excerpts, can also be created to fit specific needs.

For details, write: Special Markets, Alpha Books, 375 Hudson Street, New York, NY 10014.

Publisher: *Marie Butler-Knight*
Product Manager: *Phil Kitchel*
Senior Managing Editor: *Jennifer Chisholm*
Senior Acquisitions Editor: *Mike Sanders*
Development Editor: *Nancy D. Lewis*
Production Editor: *Billy Fields*
Copy Editor: *Ross Patty*
Illustrator: *Chris Eliopoulos*
Cover/Book Designer: *Trina Wurst*
Indexer: *Brad Herriman*
Layout/Proofreading: *John Etchison, Becky Harmon*

Contents at a Glance

Appendixes

Contents

Foreword

This book is about physics. But what is physics? Why would anyone want to learn some physics? Isn't it terribly hard and mathematical? Don't you have to be an Einstein to understand any of it? Let's talk about this.

Physics is the science that has carried out the basic research that has produced electric motors and generators, radio, radar, television, nuclear reactors, lasers, x-rays, magnetic resonance imaging, transistors, radio carbon dating, electron microscopes, quantum mechanics, superconductivity .... And physics is the science that has developed high-powered mathematical descriptions of the natural universe and hopes one day to develop one single theory or principle that will describe all observable natural phenomena.

I think there are a multitude of reasons for learning physics. It's fun, beautiful, and some would say even elegant. From a more practical viewpoint, it is a wonderful foundation to have before plunging into other sciences and technologies like chemistry, geology, astronomy, electronics, engineering of all varieties, medical technology, and so on.

Okay, so it's interesting and useful, but isn't it really hard to understand? You might find this difficult to believe, but compared to high school level molecular biology or chemistry or precalculus, it's pretty simple stuff. Eventually, the planners of high school science curricula will wake up to the idea that students should first learn physics, then chemistry, and then molecular biology instead of in the reverse order, the usual practice today. Don't be frightened by the math in this book; you worked with more complicated math in the 8th, 9th, and 10th grades.

How should you go about reading this physics book? The answer depends on how much physics you already know. I'll assume you don't know much physics. You shouldn't read physics the way you do a novel or a newspaper. It won't work. Of course, it won't damage you in any way to read rapidly or to skim through a few pages of this book, but I can guarantee you won't understand or retain much. Here is how I suggest you read.

First, start at the beginning of the book, and don't skip around. I confess, I often read biographies and magazines starting at the back or in the middle, or I just jump around, and this seems to work fine. But this won't work in physics, at least not until you learn a lot more physics. Physics is sequential. The later topics require that you know the material that came earlier in the book. So start at the beginning.

Next, read slowly and carefully. Take notes summarizing all you have read (including text, math, diagrams, and graphs). Give special attention to definitions of technical words and to general principles. You will find that preparing these notes will take you a long way down the path leading to a deep understanding of physics. Many technical words used in physics have been borrowed from everyday English. They have been given a specialized or restricted meaning when used in a physics context. Consider the following four sentences:

Max is a forceful speaker.

Amy is an energetic speaker.

Ivy is a powerful speaker.

Allison is a dynamic speaker.

In everyday language the four sentences have essentially the same meaning. But in physics each of the four nouns—force, energy, power, and dynamics—from which the adjectives in the previous sentences were derived, has meanings distinctly different from the others. So the exact meaning of key words used in physics is of great importance, and we have to be careful not to carry over from ordinary English a looser, more general meaning.

Next, make sure you work out the problems. Try your hand at solving the problems in the book without looking at the author's solutions. If you get stuck, take a peek at his solution. Make up your own problems similar to the book's, but based on situations you see in the real world. Or look in other physics books for problems. You will be amazed at how interesting and different the world looks from the perspective of the new physics you are learning. Practice, practice, practice. It's fun.

And finally, review before starting a new session; look over all the material from previous sessions. Work a few typical problems. Overlearn the material. Use either this book or your own summary notes for the review. This will give you a mastery that will help you to retain your knowledge of physics the rest of your life.

Johnnie Dennis, the author of this book and recipient of the National Teacher of the Year award, is a gifted physics teacher. The many years he spent honing his teaching skills will become readily apparent as you start your tour through his carefully crafted presentation of the world of physics. Bon voyage.

Gilbert Ford, Ph.D., Nuclear Physics, Harvard University

Vice President for Academic Affairs, Emeritus, Northwest Nazarene University

Introduction

You may have thought that physics is the name of some kind of medicine. Some of my students came to class with that idea but later decided that physics is a remedy for faulty thinking about their physical world. You are exposed to a lot of information in this book that many people never contemplate because some think that physics is too difficult for the average person. That is why the topic of physics is considered a secret meant only for the chosen few. You are not convinced of that or you would not have read this far. You are correct. Nothing is beyond your reach if, at times, you are willing to stand on tiptoe and stretch a bit.

Physics is different than any course you have ever had before. You must learn the new terms, as is true of any course, and interpret many of the terms mathematically. In this book, you start with new terms and ideas associated with familiar objects, like threatening cars speeding down the freeway on both sides of your car. You then move from a straight line into a plane to describe motion. From there you will describe the motion of an object moving in a circular path at a constant speed. These ideas prepare you for the discussion of tiny particles in solids, liquids, and gases. Later you will think about the tiny particles that move through electrical circuits to fire up your television or light your home. Eventually you will consider tiny particles of light that form images in mirrors and lenses. Finally you will consider the collisions of waves in three dimensions that form interference patterns that appear to be equally spaced black lines separated by bright regions. Or would you rather be dodging pink elephants falling from the sky?

You will be able to check your progress by working a few problems at the end of some of the chapters. Several examples are outlined for you in words, solved algebraically, and then a numerical answer is calculated. You can use the examples as models for solving the problems and check your result by looking at the answer in the appendix. You are encouraged to follow the outline for solving problems in the examples and find your own solution before checking in the book. That way you can make the several checks on your work outlined in this book, such as correct units of measurement and the correct number of significant figures. Don't be frightened of the big words because they will become clear in the context of the total picture. Prepare to get the help you need and to learn something new almost every time you read the book.

How This Book Is Organized

The book is divided into five parts:

Part 1, "Motion Near the Surface of the Earth," takes you from motion along a straight line that may be described with a linear mathematical model to motion described with a quadratic model. You see the need for a new invention called a vector to explain what happens when an object moves off a straight line into a plane. These ideas prepare you to see science as a process of vibrant change and not a collection of stagnant absolute truths.

Part 2, "When Does Something Move?" shows you that there must be a cause for motion. Not only is there a cause, you identify it as a force and measure it. You see that force can give rise to a lot of notions. The realization that there is a force of attraction between every two bodies in the universe leads to an understanding of the motions of an artificial satellite of the earth as well as natural satellites of the earth and the sun.

Part 3, "What Happens When Large Things Move?" guides you to the identification and definition of work. You'll see how work is related to kinetic energy and kinetic energy to potential energy. If you do not already know what an *erg* is, you will find out and use that word in the same way that you use the words *joule* or *watt*. You develop the concepts of power, momentum, and impulse. You will predict what happens when two objects collide, either by actually touching or interacting at a distance.

Part 4, "What Happens When Small Things Move?" explores the motions of particles in solids, liquids, and gases. You will explain what pressure is using the idea of force. Recognizing that there is a distinct difference between force and pressure, you find that the two words are not interchangeable. The discovery of a different type of energy in heat leads to an understanding of temperature that you measure in different systems of measurement. Heat influences the energy of sound and how fast it travels. Sound is found to have wavelength and frequency that enables you to explain the change in pitch of the motor of a speeding car when it passes you.

Part 5, "Tiny Things Like Electrons in Motion" takes you inside the atom by using a model to name the parts of the atom. Some parts of the atom lead to the identification of charge. Charged particles are generated and detected. Charge is associated with the electron, the proton, and the ion. Charge in motion leads to an exploration of current in an electrical circuit. From motion of charge to motion of light particles leads to the explanation of images formed by mirrors and lenses. One model of light is used until another model is explored. The new model predicts observed phenomena that the old model cannot.

Extras

The sidebars in this book offer guideposts to focus your attention by providing suggestions, definitions, and valuable information about possible difficulties to be avoided. A word of encouragement is always in order.

Plain English _____

Definitions of words used in the book to help clarify the meaning of subject matter being discussed. Look for the difference between a good definition and a description of an undefined term.

 A Johnnie's Alert _____

Sometimes serves as a warning about possible difficulties that can be avoided and sometimes offers an encouraging word. Some ideas are introduced using this vehicle.

 **Newton's Figs** _____

Suggestions and tips on how best to master an idea or to remember relationships among concepts. Ideas that are apt to turn up later are emphasized.

Physics Phun

Problems are posed for mastery of complicated notions where practice with applications may promote a deeper understanding.

Acknowledgments

The first group of individuals who inspired this book is all my former students. Many took the time to comment or make suggestions about what worked and what did not make sense in my presentations. It is my desire that you are able to recognize the characters and examples used throughout the book. Thanks to each of you and especially to those who secretly changed the spelling of words on my bulletin boards.

To my wife, Shirley Dennis, who caught typos in the manuscript and kept me on my treatment recovery diet throughout the project. To Charles and Karen, Deanna and Kevin, Maureen and Steve, and Kevin and Mary who encouraged me at every turn.

To Sarah, Caitlin and John, Nicholas and Christopher, and Summer and Seth who have urged me to do a good job because they have high expectations for a book they plan to use.

I will always be grateful to Mike Sanders and the staff at Alpha for making this project possible. Nancy Lewis is a devoted editor who helped me to gather all of the parts together in a meaningful way.

Special Thanks to the Technical Reviewer

The Complete Idiot's Guide to Physics was reviewed by an expert who double-checked the accuracy of what you'll learn here, to help us ensure that this book gives you everything you need to know about physics. Special thanks are extended to Jane Nelson.

Jane Nelson has taught high school science since 1969. She has taught physics since 1980. She has been a National Physics Teacher Trainer as part of the Physics Teaching Resource Agent program funded by the National Science Foundation and administered by the American Association of Physics Teachers for over fifteen years. She has received the Presidential Award for Science Teaching from the White House, been presented with the Disney American Teacher Award, been a Tandy Scholar, a Toyota TAPESTRY Award winner, and been inducted into the National Teachers Hall of Fame. However, when asked what her greatest accomplishment is she answers, "I raised two wonderful children and taught hundreds more how to learn without my help."

Trademarks

All terms mentioned in this book that are known to be or are suspected of being trademarks or service marks have been appropriately capitalized. Alpha Books and Penguin Group (USA) Inc. cannot attest to the accuracy of this information. Use of a term in this book should not be regarded as affecting the validity of any trademark or service mark.

Part 1

Motion Near the Surface of the Earth

These are the things you deal with every day of your life. You dodge them on the freeway and hope that you do not step on them at night when you walk across the floor to turn on the lights. All objects like these move, have speed, and sometimes really change their speed in a hurry. This part will help you better understand what it means to move. You will learn that sometimes it makes a difference as to which direction you want to go. You also find that ten plus ten is not absolutely twenty—it could be zero! Even though the objects discussed are everyday kinds of things, you may find that some brand new ideas are needed to describe how they move.

Chapter 1

Speaking Physics

In This Chapter

- ◆ The language of physics
- ◆ Units of measurement
- ◆ Significant figures
- ◆ Scientific notation
- ◆ Calculating using rules of scientific notation

When you visit a new part of the world, it would be helpful if you were able to speak the language of the natives in order to understand their culture. Well, the world of physics is revealed through mathematics. In fact, a lot of people think erroneously that knowledge of physics involves memorizing the correct formula for each new situation. If that were the case, an individual would have a head full of formulas looking for a place to use them in physics. Actually, to know physics you need a clear head for thinking critically and just enough mathematics to speak the language of the subject. That is why you need this book … so that I can guide you through difficult ideas and help you to feel comfortable with a difficult subject.

When I first studied physics in high school, I felt lost because there were so many strange and apparently unrelated ideas. No one had bothered to explain which mathematical ideas were necessary for me to be successful in physics. If you have ever had that experience, I am here to help you avoid those feelings of insecurity.

Together we will revisit some of those areas of mathematics that I am sure were once at the tip of your tongue. In the event there are some of those ideas that you did not master the first time, I will guide you step by step through the key notions you will need to speak physics. You do not have to be a genius in mathematics to enjoy physics, nor do you have to be a nerd to want to learn everything you can to understand physics. This book is comprehensive and will help you with all topics of physics you may ever tackle. All that you need is the interest in the subject and the willingness to do a minimum amount of practice. When you apply the language enough so that you are able to use it, you will have almost as much fun as eating a cold piece of watermelon while enjoying a hot southern evening.

This chapter contains information that you will refer to again and again as you read this book. You do not need to master every idea presented here but you should know that you have it at your fingertips anytime you need it.

Making Measurements

We will begin with the systems of measurement used to measure physical quantities. The metric system used includes the CGS and MKS units of measurement. The MKS is shorthand for meter, kilogram, and second. Similarly, CGS is a short way of naming centimeter, gram, and second. Another system that is used, the British System, is called the FPS system; FPS is just a short way of labeling foot, pound, and second. Fortunately, the MKS system is so widely used that it is included as part of SI, or International System of Units. We will use all units to practice some mathematics soon. Learn the units of measurement in each of CGS, MKS, and FPS systems. From now on in this book the three fundamental quantities of physics will be measured in these units.

The letters CGS, MKS, and FPS in the titles of those systems are used for the three *fundamental quantities* of physics that we think about early in the study of physics. These quantities are *mass*, *space*, and *time*. I will list them in a table here to help you as we consider these different systems of measurement. The table will be something that you will want to refer to throughout your reading of this book and you will find that the table will be extended to include *derived quantities* which are also very important in our effort to understand the language of physics.

Units of Measurement of Physical Quantities

Fundamental Quantities	MKS	CGS	FPS
Time	s	s	s
Space	m	cm	ft
Matter (mass)	kg	g	slug

As you can see, the fundamental quantity of time is measured in the same units in all systems. You will see later that there is a nice mathematical relationship between units in the CGS system and the MKS system. But the FPS system is something else! A slug! You think of a slick, slimy, crawly creature that steals bites out of your vine-ripening tomatoes. Here the slug is the unit of measure of mass in the FPS system. Is it any wonder that students of physics prefer SI units? Unfortunately, the FPS system is the system used widely in the United States even though efforts are still being made to change to SI units. Hmmm, did you notice that the P for pound is not even in the table? What is going on? Well, you are learning to speak physics when you notice things like that. We will find an explanation for that in a later section. Actually, the simple reason is that the pound is not a fundamental measure of matter. Hold on to the thought, though. We will come back to it to satisfy your curiosity in Chapter 5.

Plain English

Fundamental quantities are the building blocks for the foundation of physics. Time, space, and matter are the quantities required to study that area of physics called Mechanics. **Derived quantities** are all of those physical quantities that are defined using two or more fundamental quantities like meters/second or one fundamental quantity used more than once like meters cubed.

Recording Measurements and Significant Figures

We should look at a couple of other ideas so that we can use this language to play the game of physics at least on an introductory level for the moment. First consider *significant figures*. A lot of playing physics involves experimentation where measurements are made. When you make a measurement you should record and report the best observation you can make. In order to do that, you should record every digit that you are sure of, given the measuring instrument that you are using, plus one more digit that you estimate. That way you and anyone who reads your report will know that any error that may occur will be recorded in that last digit you write down.

Even when you take all of this care, not all digits are significant when you record measurements using the conventions of writing numerals. I will list some rules here that you can follow to make sure your records communicate the same information to anyone who may read your work. Keep this list handy until you know the rules:

1. All nonzero digits are significant. Example: In the number 134 m, there are three significant digits.

2. All zeros occurring between nonzero digits are significant. Example: In the number 104 m, there are three significant digits.

3. All zeros occurring beyond nonzero digits in a decimal fraction are significant. Example: In the number 0.1340 m, there are four significant digits. They are the last four digits.

4. Zeros used to locate the decimal point are not significant. Example: In the number 0.0104 m, there are only three significant digits. They are the last three digits. Also, in the number 25000 mi, there are two significant digits because the three zeros serve only to locate the decimal point.

These rules apply a) whenever you make a measurement; or b) when the statement of a problem implies a measurement. Notice that a unit of measurement will be a part of such a record. So whenever you see a numeral with a unit of measurement you will know that significant figures must be observed. That is true with one exception. If the numeral with units results from a *defining equation*, then all digits are significant. It should also be noted that all numerals without units of measurement contain digits all of which are significant.

You are already familiar with defining equations but you may not have called them by that name. Consider some of those equations for a moment because using them properly will be of unbelievable value to you later on: 1 hr = 60 min, 1 min = 60 s, 1 mi = 5280 ft, 1 kg = 1000 g, 1 m = 100 cm. All of these are familiar to you, I am sure. Later we will talk about *unit analysis* and at that time I will show you how to use these equations to define unity. Then you will be able to change comfortably from one measurement to another. There are many other defining equations; I will introduce some of them to you and demonstrate how they are used at the proper time. That is an awful lot to remember, so keep your list handy. Some examples of each of these rules will help. Then you can have some Physics Phun.

Physics Phun

How many significant figures are recorded in each of the following: (1) 2500 km, (2) 3.14, (3) 0.005 m, (4) 3.100 m, (5) 300,000,000 m/s.

Plain English

Significant figures are those digits an experimenter records that he or she is sure of plus one very last digit that is doubtful. A **defining equation** is a statement of a relationship between two units of measure. **Unit analysis** is the process changing from one unit of measure to a larger or smaller unit of measure without changing the value of the measured quantity. You may do this by defining a disguise of unity (one) in terms of the units you wish to change.

Counting Significant Figures

Recorded Measurement	Number of Significant Figures
102.33 m	5 (by rules 1 and 2)
10.0500 s	6 (by rules 1, 2, and 3)
92,000,000 mi	2 (by rules 1 and 4)

Standardizing Notation

The other notion is *scientific notation*. You probably recall from science classes and newspaper accounts that physics involves some numbers that are very large and some that are very small. That means that we will have to become proficient at writing lots of zeros to represent those numbers unless there is a shorthand method of doing that for us. Scientific notation is just the shorthand we need for writing numbers like three hundred million and six hundred-thousandths. Even writing these numbers in words is confusing but our shorthand will remove any doubts you might have about that. In addition, we will quickly and easily record the numbers with the correct number of significant figures.

A number is written in scientific notation in the following way: First remember that the pattern the number in scientific notation must fit is the following $\_.\_ \times 10^n$ where the first blank is for a digit from 1 to 9 and the second blank is for all remaining significant figures in the original notation multiplied by the power of ten that makes the number fitting that pattern equivalent to the original number.

Plain English

Scientific notation is a method of writing observations in a special pattern as follows: $\_.\_ \times 10^n$ where the first blank is a digit from 1 through 9 inclusive and the remaining blanks are for all remaining significant figures. The power of ten is chosen as a multiple to make the number in this form equivalent to the observed number.

For example, the measurement 92,000,000 mi in scientific notation will be 9.2×10^7 mi. The measurement can be read ninety-two million miles and the scientific notation nine point two times ten million or ninety-two million. Isn't that remarkable? Any doubt about which type of notation is better? If there is, try this number 602,000,000,000,000,000,000,000. Would it not be easier to write 6.02×10^{23}? How about a few more examples? We will write these measurements in scientific notation:

Examples of Writing Numbers in Scientific Notation

Original Number	Written in Scientific Notation
250,000 mi	2.5×10^5 mi
0.003 m	3×10^{-3} m
0.005200 g	5.2×10^{-3} g
2.0500 km	2.0500×10^0 km

You will not see the case listed in the last entry in the table very often but you should know how to handle it when it arises. You know that $10^0 = 1$.

Physics Phun
Write the following numbers in scientific notation: (1) 250,000 mi, (2) 4,830,000 m, (3) 23,000,000 s, (4) 0.000876 kg, (5) 1.0003 cm.

Newton's Figs

The number 6.02×10^{23} is Avogadro's Number, the number of things in a mole. Can you imagine a mole of baseballs? The number 9.2×10^7 mi is the distance from the earth to the sun. A mole is a unit of quantity and is Avogadro's Number of objects. The large numbers listed here serve to illustrate the importance of scientific notation unless you are fond of writing zeros.

Speaking the Language

Now that you have significant figures and scientific notation fresh in your mind, I will review with you methods of calculating some physical quantities using given measurements. In order to keep track of the precision of our calculations, we must abide by some rules listed here:

- ◆ When adding or subtracting, round off the terms of the sum or difference to the precision of the least precise term in the sum or difference then carry out the operation.

- ◆ When multiplying or dividing, carry out the operation and then round off your answer to have only as many significant figures as the factor in the product or quotient having the least number of significant figures.

Why Ask Why?

Before making those calculations, I will explain a few things to you concerning the task ahead. For forty years I was fortunate to have taught physics to high school students. Almost every year students would ask me, "Why are we going to all of this trouble?" That always pleased me because it gave me a chance to explain that there are good reasons for using scientific notation and significant figures. Otherwise, I would not have wasted their time in the classroom.

I will not waste your time with facts or processes that are useless to your understanding the ideas in this book. So this is the beginning of an answer to a difficult question, "Why?" You will see as you read this book, answering the question "Why?" in physics is not always possible. For example: Why does the earth spin on its axis? I do not know the answer and no one has ever given me an answer other than saying, "The earth does spin on its axis." That does not mean that you should not ask the question, but you should not be disappointed if you do not get an answer. Someday you may be the person who will give the world a good answer to such questions.

We are going to all this trouble because in science in general and in physics in particular we try to tell the best truth possible. You can find a discussion of truth in a lot of sources but probably all will agree with the following observations. In science we do not look for absolute truth, that is, something that is true now, always has been, and always will be true. If you are ever made aware of a scientist saying, "It is absolutely true that this or that scientific event will happen," be watchful because you are about to be scammed! If you want to study absolute truth, consult with your religious advisor or a trusted philosopher.

Goals = Probable Truths

Even the mathematician does not seek absolute truth. A mathematician's goal is "relative" truth. He does assume certain things to be true absolutely, such as axioms and postulates, but he proves theorems to be true relative to the truth of those ideas he has assumed to be true absolutely. The scientist's objective is probable truths. That does not mean maybe a scientific law is true and maybe not. Men and women who make the best measurements possible establish probable truths. They first use their senses to make observations. Then they use their intellect to formulate a hypothesis and to test their hypothesis in the laboratory. After all of that activity, they report the best truth that they can find guided by the theory of probability.

Physics is exciting when you consider the efforts men and women have made over the years to construct such an outstanding body of knowledge based on probable truth. That body of knowledge is the discipline of physics that I am sharing with you in this book. The beauty of physics is that scientific explanations may and do change as more studies are made. Physicists would be neurotic or crazy if they believed that their current body of knowledge is absolutely true. If scientists thought they had the absolute truth about the effects of radiation 30 years ago, they would not have found how to use it to treat some cancers, but they have.

I know that to be true because doctors recently cured me of cancer. They used a radiation treatment developed for medicine by physicians in cooperation with physicists. If scientists believed that their knowledge pool is absolutely true, we would never have gone to the moon, we would not have satellites for television and cell phones, but we do have all that and more. None of these things would have happened had we believed scientific statements to be absolute, truths such as "Matter can neither be created nor destroyed," a statement that was in my high school physics book in 1951.

So you can see that we go to all the trouble of observing significant figures and using scientific notation to try to tell the best truth we can as we enjoy the fun of exploring physics. Numbers that are measured are only as good as the instruments used to measure them and the person used to read them. If we record all the numbers that are shown on the face of a calculator after it has multiplied two numbers, we are implying more precision in the instruments used to take the measurements than is true. If this is your first exposure to physics or if you are revisiting the difficult subject using this new approach, keep these ideas in mind as you read more and discuss physics.

Calculations Using Significant Figures

Now that I have given you a reason why we use certain ideas, I will show you how to calculate some quantities using significant figures to point out how that notion helps us to tell the best truth possible. Not absolute truth, but probable truth; also since you know that mathematics is the language of physics, you should know that we are borrowing some ideas from mathematicians that involve relative truth.

Suppose you want to add and subtract some measurements of distance. Arrange the sum or difference so that the decimal points are in a vertical line as in the following examples: The first column for each operation below is the original set of terms to be combined. The second column in each case is the set of terms rounded to the precision of the least precise term in the sum or difference. When rounding numbers use the following rules:

- If the number being rounded off is greater than five, then drop the number and increase the preceding number by one.

- If the number being rounded off is less than five, then drop the number and leave the preceding number unchanged.

- If the number being rounded off is exactly five and the number preceding is odd, then drop the five and increase the preceding number by one. If the number preceding the five is even, then drop the five and leave the preceding number unchanged.

Add these measurements:

283.6 cm	283.6 cm
34.621 cm	34.6 cm
91.25 cm	91.2 cm
8.36 cm	8.4 cm
	417.8 cm

Subtract these measurements:

478.348 m	478.3 m
332.1 m	332.1 m
	146.2 m

Now to multiply and divide use the same rounding procedures but follow the rules for calculating using scientific notation.

Multiply these measurements:

151 ft
46 ft
906
604
6900 ft^2

Divide these measurements:

$$\frac{480.6m^2}{47.8m} = 10.1 \text{ m}$$

Notice that the units of measurement behave just like numerals in calculations, that is, ft × ft= ft² and m²/m = m. You will get a lot of practice with this later, but I wanted to focus your attention on this important property.

Now it is your turn with some Physics Phun.

Physics Phun

Add the measurements: 36.98 cm, 905.6 cm, 8.64 cm. Subtract the smaller of these two measurements from the larger: 348.256 ft, 24.3 ft. Find the product of 671.89 m and 53.4 m. Divide the larger measurement value by the smaller: 3258.25 cm², 42.5 cm.

The Least You Need to Know

- Recognize that the three fundamental quantities of physics are space, mass, and time.

- Use significant figures to record observations.

- Write numbers in scientific notation.

- Explain the nature of truth sought by scientists.

- Identify the fundamental units of measurement in the MKS, CGS, and FPS systems.

Chapter

2

On the Move Again

In This Chapter

- ◆ Talking about speed
- ◆ Check the speedometer
- ◆ Changing the motion
- ◆ Calculating velocity and speed

Your favorite NASCAR driver heading his car down the straight portion of the track causes his car to have velocity; and if it has velocity, it has speed. In fact, as long as he is traveling in one direction in a straight line, his speed and velocity are the same. If he speeds up, his car is accelerating.

You will explore the relationships among the ideas of velocity, speed, and acceleration in this chapter. You will be able to describe the motion of an object—whether it is speeding up or traveling with a constant speed. The beginning trip will not be as exciting as the racecar but the ideas are rewarding enough for the ride.

Understanding Speed

Because you are reading this book, it is obvious that you are not a couch potato. However, I suspect that you have observed a few of those creatures in your life. Suppose you see a couch potato sitting at one end of the couch, and a short time later he is sitting at the other end. Obviously, he had to *move* in order for the two observations to make sense. Call the first place he was sitting his initial position and the last place he was sitting his final position. By thinking physics, it is reasonable for you to say that he has experienced a change in position.

Plain English

An object is said to **move** when it experiences a change in position.

Not only is there a change in position, but also during that process his body was in motion, which means that it had *speed*. The amount of speed depends on how long it took to make the change in position, that is, the amount of time required to go from his initial position to his final position on the couch, based on the path he took as he went from one position to the other.

Motion Discussion

We now have the principal elements for a discussion of motion. Not much is known about his speed so far. We will begin by assuming that several speeds were involved in accomplishing the change in position.

It took a lot of words to describe this interesting situation, but the language of physics will show how simple such a discussion can be. Algebra plays a very important role in helping you to understand these concepts of physics. Don't worry if you do not remember algebra. I will lead you through a step-by-step solution in which algebra is used.

Consider the following symbols, which are used to represent some concepts instead of words in this discussion:

- x_0 = initial position, or position when we begin measuring time

- x_f = final position, or position at the end of measured time

- v_0 = initial speed, or velocity when we begin measuring time

- v_f = final velocity, or speed at the end of measured time

- $\bar{v}$ = average velocity, or speed over the time interval

- t_0 = initial time, or the reading on a stopwatch to begin measuring time

- t_f = final time, or the reading on a stopwatch at the end of the measured time interval

- Δ = symbol used to represent change

We use x to represent the position at any time. The change in position will be $\Delta x = x_f - x_0$. The change in time can be expressed as $\Delta t = t_f - t_0$. Similarly, if v

represents the velocity at any time then $v_f - v_0 = \Delta v$, the change in velocity. The *average velocity* is the change in position divided by the change in time: in symbols,

$\bar{v} = \dfrac{\Delta x}{\Delta t}$, which is $\bar{v} = \dfrac{x_f - x_0}{t_f - t_0}$.

Imagine that a straight line can represent the couch. Distances x can be marked off along the line from one end of the couch to the other. Suppose we measure all distances relative to the first place the couch potato was sitting so that $x_0 = 0$ and $x_f = x$. Furthermore, suppose that we started the stopwatch when he first moved from the initial position and stopped it when he reached his final position so that $t_0 = 0$ and $t_f = t$. That means that $\bar{v} = \dfrac{x_f - x_0}{t_f - t_0} = \dfrac{x - 0}{t - 0} = \dfrac{x}{t}$.

In other words, the *average velocity* is equal to the length of the couch, x, divided by the time, t, required for him to travel that length. Therefore, from $\bar{v} = \dfrac{x}{t}$ the expression $x = \bar{v}t$ represents the position at any time t assuming that the couch potato is always traveling with the same motion. In words, the position or distance traveled at the end of a time interval is calculated by multiplying the average speed during that time interval by the time of the interval.

Plain English

The **average velocity** is the change in position of an object divided by the change in time.

Using Algebra to Calculate Velocity

Algebra enables us to communicate several ideas clearly and quickly with a few simple mathematical statements. Consider the following example.

Suppose you drive your car to school from home. Your route takes you along neighborhood streets to the interstate. You drive down the interstate for a few minutes. You then travel along neighborhood streets to the school parking lot where you park. The school is 15 miles from home and you make the trip in 3.0 minutes by driving within the speed limit of course. What is the average speed of your car for the short journey?

Solve any physics problem by first analyzing it into its parts and specifying properties of each part. In this case, we are given: (1) the distance in miles with two significant figures and (2) the time in minutes with two significant figures. Rarely do you see speed expressed in terms of minutes; it can be, but we use hours because that is a common unit used in the United States to indicate speed. From a defining equation, express the time as 5.0×10^{-1} hr or ½ hour to two significant figures. Secondly, we synthesize the parts with our definition of average velocity and solve the problem. That is,

$\bar{v} = \dfrac{x}{t} = \dfrac{15mi}{5.0 \times 10^{-1} hr} = 3.0 \times 10^1 \dfrac{mi}{hr}$.

Notice that the units are written as a fraction and are read miles per hour in this case. Soon I will guide you through the details of determining the correct units for a given situation by using unit analysis. I suspect that you knew immediately what the units would be for the answer to this problem. It is a familiar problem and one that you probably solve daily without even thinking. That is good. It is my intent to help you to realize that you know a lot of physics stuff already. This book will help you to use that knowledge to guide you in your desire to understand more physics.

Describing Uniform Motion

I admit the use of a ploy to fix your attention on movement and speed. That is, I used your knowledge of motion to guide the discussion of average velocity. Somewhere in your background you know that distance is equal to rate times time. That idea was used to introduce the relationship between average speed and distance. You were reminded that the speed is not the same throughout the trip of the couch potato. It might have been zero at times. From your observations of his behavior, it probably is true that he moved from one end of the couch to the other in a series of rest stops.

After calculating the average speed, I described the motion as if the speed was the same throughout the journey. Averages have that property of covering up a lot of irregularities. Averages are very much like the touch-up paint job you use on your car when there is a scratch. The paint from your touch-up makes the surface look nice and smooth as if it is the original paint. Similarly, average speed is nice and smooth and covers up any changes that might have occurred during the motion.

We treated average speed as if it was the same and the results we achieved are fine. What happens if the speed is constant? Well, as it turns out, one of the simplest types of motion to describe is *uniform motion*. An object traveling with uniform motion is traveling with a constant speed. Unlike the motion we reviewed in the last section, the speed does not change, it is always the same. Except for the fact that the speed is now constant, everything we said about motion in the last section is true of uniform motion. Here is a list of the symbols we will use to describe uniform motion:

Plain English

Uniform motion is motion that is characterized by a constant velocity or speed. Whenever a situation in physics states or implies uniform motion, then you will know that the speed is constant. If the speed or velocity is constant, then you know that the object has uniform motion.

- ◆ v = the constant velocity
- ◆ x = the distance traveled
- ◆ t = the time the object is in motion

Uniform motion is described simply as $x = vt$. Since the velocity is constant, the distance traveled is found by calculating the product of the average velocity and the time. The average velocity is the same as the velocity at any time because the velocity is constant.

Suppose you drive along the interstate, set your cruise on $80\frac{km}{hr}$, and continue on your way for 30 min. How far will you travel? You probably know that 40 km is the correct answer. You are given a constant speed of $80\frac{km}{hr}$ and a time of 30 min. both having one significant figure. That means that your car is traveling with uniform motion and since $x = vt$, $80\frac{km}{hr} \times \frac{1}{2}hr = 40km$.

"Not so fast" you say? Well, the answer has one significant figure and the unit of measure is the kilometer. Remember where that unit came from? Maybe it is time for a discussion of defining equations and unit analysis I promised earlier. All right. Hold on to your hats. Here goes!

The emphasis here is not on significant figures and scientific notation. We have reviewed those ideas and need only to practice to become proficient using them. We now consider the units of measure and the units of our final answer. Using the information from the last section, here is the detailed analysis of the units. Since $x = vt$, $x = 80\frac{km}{hr} \times \frac{1}{2}hr = 40km$. The question is, "How do we get km?" We were given that the constant speed is $80\frac{km}{hr}$ and the time is 30 min and we know that we want distance in km in this problem. By using the defining equation 60 min = 1 hour, we can define unity to be disguised as $\frac{hr}{min}$.

A Johnnie's Alert

Always observe significant figures when calculating results that include measured quantities. Always use scientific notation if measurements of physical quantities are very large or very small.

That is, dividing both sides of that defining equation by 60 min we get $1 = \frac{1hr}{60\,min}$. We can multiply any quantity by 1 without changing the value of the quantity. We just change its appearance. Therefore, $80\frac{km}{hr} \times \frac{1hr}{60\,min} = \frac{80km}{60\,min}$ because $\frac{hr}{hr} = 1$. Remember the units behave like any other rational numbers. When we multiply $\frac{80km}{60\,min} \times 30min$, we get 40 km because $\frac{30\,min}{60\,min} = \frac{1}{2}$ and $80km \times \frac{1}{2} = 40km$.

Using the defining equations and unit analysis we will be able to express our answers in any units of measure we desire. You may design unity (1) to have any disguise that may be helpful to your solution by using legal mathematical operations on a defining equation. An illegal mathematical operation is division by zero. As you know, division by zero in undefined. Take a look at a couple of other examples.

Express 6.0×10^1 mi/hr in ft/s. Use the following defining equations: 5280 ft = 1 mi, 60 s = 1 min, 60 min = 1 hr. Disguise unity (1) using each of those equations and write the result: $6.0 \times 10^1 \dfrac{mi}{hr} \times 5280 \dfrac{ft}{mi} \times \dfrac{1hr}{60 \min} \times \dfrac{1 \min}{60s}$, that is, $6.0 \times 10^1 \dfrac{mi}{hr} \times 1 \times 1 \times 1$, which is still $6.0 \times 10^1 \dfrac{mi}{hr}$, but because of the different disguises of unity (1), the final answer will be $88 \dfrac{ft}{s}$. If you were driving on the interstate at this speed on a rainy day, you could easily wind up in a ditch with a big chunk of metal for a car.

Express 1 year in seconds. Use the defining equations: 1 yr = 365 days, 1 day = 24 hr, 1 hr = 60 min, and 1 min = 60 s. Then disguise unity from each of those equations and write the following: $1yr \times 365 \dfrac{da}{yr} \times 24 \dfrac{hr}{da} \times 60 \dfrac{\min}{hr} \times 60 \dfrac{s}{\min}$, which is another way of writing $1yr \times 1 \times 1 \times 1 \times 1$ (which is still 1 yr) but the disguises of unity result in an answer of 3.1536×10^7 sec, the number of seconds in a year.

Remember that all of the digits are significant in defining equations and 1 yr was not measured. Do you agree that this process of unit analysis is a nice, easy way of keeping track of units of measurement? All you need are the defining equations that can be found in many sources. You will not have a lengthy list of them here since they are presented throughout this book.

Physics Phun

Use unit analysis and appropriate defining equations to express: (1) 9.2×10^7 mi in meters, (2) 1 day in seconds. Hint: Remember that all digits are significant in defining equations such as 1 day=24 hours. Only measured quantities limit the number of significant figures in your calculations.

Understanding Acceleration

Have you seen a guy with a green 72 Dodge Dart leave the parking lot so fast that sparks fly from his bumper as it scrapes the surface of the road? He says, "It will do sixty in a quarter!" He is telling us that the car can speed up. Since he has brakes good enough to stop at the intersection, the car can also slow down. In physics speak, if a car speeds up, the car is said to accelerate. If it slows down, it is said to have negative acceleration or it decelerates. *Acceleration* is expressed in terms of velocity.

The car's velocity is not as simple as the velocity from the last section because in that case the velocity was constant for uniform motion. Obviously, you are observing something much different when you see the Dart go from stand-still to at least 30 mi/hr out of the parking lot.

Plain English _____

Acceleration is the rate of change of velocity or the change in velocity divided by the time for that change to take place.

The speeding up or slowing down implies a change in speed. We need to identify some symbols necessary to discuss acceleration:

- a = acceleration

- v_f = final velocity or the velocity at the end of the time interval

- v_0 = initial velocity or velocity at the beginning of the time interval

- t_0 = initial time or time at the beginning of the interval

- t_f = final time or time at the end of the interval

Using this information we can find Δv and Δt then define acceleration as follows:
$a = \dfrac{\Delta v}{\Delta t} = \dfrac{v_f - v_0}{t_f - t_0}$, that is, acceleration is the change in velocity divided by the change in time or the rate of change of velocity.

Whenever a new concept of physics is presented, you should check the units of measurement immediately. The definition of acceleration and the MKS system are used here to identify the units of acceleration in that system. Since $a = \dfrac{\Delta v}{\Delta t}$, Δv will have $\dfrac{m}{s}$ as units and Δt is measured in s. Now $\dfrac{\Delta v}{\Delta t}$ will have units of $\dfrac{\frac{m}{s}}{\frac{s}{1}} = \dfrac{m}{s} \times \dfrac{1}{s} = \dfrac{m}{s^2}$ and is read meters per square second or meters per second per second.

Don't try to visualize a square second. There is no such physical quantity. Since units of measurement behave like any other rational numbers, when the division is performed the units that result are meters per square second. That means simply that the speed in meters per second changes every second. Did you notice that we are working with quantities that are not fundamental quantities? It is high time to expand the table of units of measurement.

Units of Measurement of Physical Quantities

Derived Quantities	MKS	CGS	FPS
Speed	m/s	cm/s	ft/s
Acceleration	$\dfrac{m}{s^2}$	$\dfrac{cm}{s^2}$	$\dfrac{ft}{s^2}$

There are many other derived quantities, and I will either continue constructing additional entries or ask you to continue expanding this table for yourself.

Describing Uniformly Accelerated Motion

Recall that we were able to describe the motion of the couch potato by using his average velocity. We will use that same idea here but we must keep in mind that we want to describe the motion of an object that has a constant acceleration. That makes this a very different discussion. We use the same symbols as we have used before to describe motion. There are three basic ideas you need to develop the solution to any problem concerning an object that has constant acceleration. Some of the algebraic solutions you will want to be able to construct are outlined here as well.

Any time that you consider a problem that states or implies an object moves with constant acceleration these three ideas apply:

- $x = \bar{v}t$ Distance is equal to the average velocity or speed times the time.
- $\bar{v} = \dfrac{v_f + v_0}{2}$ Definition of average velocity where v_f and v_0 are given or implied. v_f is the velocity at the end of the interval of time and v_0 is the velocity at the beginning of the time interval.
- $a = \dfrac{v_f - v_0}{t}$ The definition of acceleration where is the length of the time interval.

Look at this example. You are in a car that was initially at rest but is now traveling with acceleration a and continues to accelerate until it reaches a final speed of v_f. How far did you travel? Use the procedure for solving a physics problem mentioned earlier. Solve the problem first in general (algebraically) then substitute given information for this particular case. In general $x = \bar{v}t$, that is, distance equals average velocity times the time. Use the definition of $\bar{v}$ to find that $x = \left(\dfrac{v_f + v_0}{2}\right)t$. Using the

definition of acceleration we find that $x = \left(\dfrac{v_f + v_0}{2}\right)\left(\dfrac{v_f - v_0}{a}\right) = \left(\dfrac{v_f^2 - v_0^2}{2a}\right)$ for the general solution. The particular solution is obtained by substituting the information implied in the statement of the problem, $v_0 = 0$, and finding that $x = \dfrac{v_f^{\,2}}{2a}$. We still have a solution in algebraic symbols but the beauty of this solution is that you have solved all problems of this type.

Physics Phun

An object has an initial velocity v_0 and an acceleration a. How far does it travel in a time t? Hint: If $a = 6.0\,\dfrac{ft}{s^2}$, $v_0 = 6.0\,\dfrac{ft}{s}$, and $t = 1.0 \times 10^1\,s$, then $x = 360\,ft$.

Remember that this is a discussion of motion in one direction along a straight line. Velocity and speed are closely related and may be used interchangebly under these conditions. You describe uniform motion by the simple statement $x = vt$ and uniformly accelerated motion is described using three ideas: $x = \bar{v}t$, $a = \dfrac{v_f - v_0}{t}$, and $\bar{v} = \dfrac{v_f + v_0}{2}$.

The Least You Need to Know

◆ Write the definition of average speed given distance and time.

◆ Describe uniform motion in terms of distance, speed, and time.

◆ Describe uniformly accelerated motion using three equations.

◆ Use defining equations and unit analysis to account for changes in units of measurement.

◆ Use the equations for describing uniform motion to solve motion problems algebraically and arithmetically.

Direction Does Make a Difference

In This Chapter

- ◆ Things that need direction and those that don't

- ◆ The rules of the game

- ◆ The roles parts play

- ◆ A close relative of distance

- ◆ The identity of speed revealed

If you have dealt with vectors before, you know what strange creatures we are going to be discussing here. If this is your first time through, be patient and play the game. The first time I heard about vectors I thought my professor was joking. Maybe these things are strange because strange individuals created them; physicists created this idea, and mathematicians borrowed it and generalized the notion.

After mastering this chapter on vectors, you will be able to row your canoe directly across a swift stream. You will know that in order to go straight across you will have to keep the canoe pointing upstream just the right amount and row like crazy. If you are ready, we will enter the V zone.

Understanding Vectors and Scalars

A *vector* is a quantity that has magnitude, direction, and obeys a law of combination. The law of combination is the commutative law of addition. That is, 4+3=3+4. In other words, changing the order of adding terms yields the same result. Vectors must have all three properties. If a quantity is missing any of these properties, it is not a vector.

Plain English

A **vector** is a quantity that has magnitude, direction, and obeys a law of combination. The magnitude is the size measured in proper units of measurement. A **scalar** is a quantity that has magnitude only. Any units of measurement are included when we refer to magnitude.

Are there quantities that are not vectors? Yes. You are already familiar with one set of such creatures; you have used them all your life. They are *scalar* quantities. You probably do not recognize the name but you will recognize the quantities. A scalar is a quantity that has magnitude only. The magnitude includes the units of measurement, of course. Some examples are distance, speed, and time, to mention just a few. We discussed all of these in an earlier section. When a police officer gives you a speeding ticket, he is not concerned with the direction you are going, unless of course you are traveling the wrong way on a one-way street.

We can write *25 km* and know immediately that is a measurement of distance, a scalar quantity, because there was no reference to a direction. How do we represent a vector quantity? We use an arrow to represent a vector. That is, an arrow is to a vector quantity what the numeral 3 is to the number three. You may not have ever made that distinction before, but 3 is the symbol used to represent an abstract idea we refer to as three. The numeral 3 is unique in that it represents one and only one number. An arrow of given length and direction represents a unique vector quantity.

Recognizing and Dealing with Vectors

An arrow is a plane figure. In a given plane there can be many arrows having the same length and direction; all such arrows represent vectors that are said to be equal. It is good for you to see some of these odd birds and see how they differ from scalars. In Figure 3.1, I have given you a representation of two vectors. Included also is a label for each. The symbols $\vec{A}$ and $\vec{B}$ are used to refer to the arrows representing the corresponding vectors so that we may discuss the vectors without drawing the arrows. In addition, these two vectors are equal because they have the same length and the same direction. In terms of the new symbols, we say $\vec{A} = \vec{B}$.

The head of the arrow gives the direction. The arrowhead implies an angle between the vector and a horizontal line as well as between the vector and a vertical line. I will give you more details about measuring these angles in the next section of this chapter. The length of the vector is measured from the tip of the arrow to the other end, called the foot. The length of the vector is the magnitude that includes the proper units of measurement of the vector.

Newton's Figs _____

Two vectors are *equal* if they have equal magnitudes and the same direction. Equality of vectors is very special in that two vectors are equal if and only if they have the same magnitude and the same direction.

Figure 3.1

Vectors $\vec{A}$ and $\vec{B}$ demonstrate equal vectors. $\vec{A}$ and $\vec{B}$ are algebraic symbols used to name the arrows that represent vectors.

Vector Addition, a New Way of Adding

Now that we have established what we mean by equal vectors, we will look at the law of combination. Consider the two vectors in Figure 3.2.

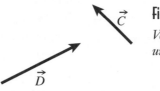

Figure 3.2

Vectors $\vec{D}$ and $\vec{C}$ are unequal vectors.

Obviously, vectors $\vec{D}$ and $\vec{C}$ are not equal. They have neither the same direction nor the same *magnitude*. I use these vectors to demonstrate a method for adding vectors. What is true about adding these two vectors is true for adding any two vectors, even equal vectors.

Plain English

Magnitude is the size of the vector in the proper units of measurement. It is the length of the arrow representing the vector. The length can be drawn to scale to represent the magnitude of any vector.

Follow these rules to add the vectors in Figure 3.2: (1) construct vector $\vec{C}$ at some convenient place in the plane, (2) construct vector $\vec{D}$ with its foot exactly at the tip of the head of vector $\vec{C}$, (3) draw a new vector from the foot of $\vec{C}$ to the head of $\vec{D}$. Call the new vector $\vec{R}$, which is called the sum of $\vec{D}$ and $\vec{C}$. Using the new algebra of vectors that we are developing, we can state for this case $\vec{C} + \vec{D} = \vec{R}$.

Figure 3.3

The vector sum demonstrates a method of adding vectors.

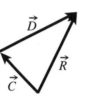

Newton's Figs

Remember the rules for adding vectors. These rules will apply any time you add two or more vectors together.

In Figure 3.4, instead of starting with $\vec{C}$ I begin with $\vec{D}$ and add the same two vectors. That is, construct $\vec{D}$ first and then place the foot of $\vec{C}$ exactly at the head of $\vec{D}$. Draw the new vector from the foot of $\vec{D}$ to the head of $\vec{C}$ and compare it with the vector $\vec{R}$ that we found in the previous sum. No tricks, no magic, but they are equal, exactly the same vector. That is, it has been demonstrated that $\vec{C} + \vec{D} = \vec{D} + \vec{C}$.

Any time you combine two vectors, use these rules and find that the law of combination is obeyed. Furthermore, you can add any number of vectors if you continue to join them together foot to head, foot to head. If you combine more than two vectors, the vector sum, the *resultant vector* $\vec{R}$, is obtained by joining the foot of the first vector to the head of the last vector drawn. Refer to Figure 3.5 for an example of combining more than two vectors.

Plain English

The **resultant vector** or just the resultant is the name assigned to the vector representing the sum of two or more vectors.

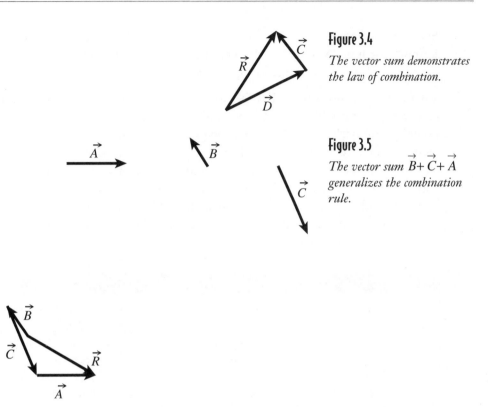

Figure 3.4

The vector sum demonstrates the law of combination.

Figure 3.5

The vector sum $\vec{B} + \vec{C} + \vec{A}$ generalizes the combination rule.

I use three different vectors to mix some variety into this discussion. Otherwise, these vector monkeys may get a little on the feisty side. I chose the order $\vec{B} + \vec{C} + \vec{A}$, but you know that any order for the terms may be used and the result is the same for the given vectors. Remember, in order to find the sum, always draw the resultant vector from the foot of the first vector you construct to the head of the last vector in your sum. I indicated that vector with the symbol $\vec{R}$. You may use any symbol you please to denote the vector sum.

Did you notice that I changed neither the magnitude nor the direction of any of the vectors in Figure 3.5 when I added them together? When you practice adding vectors, you must be sure to preserve magnitude and direction of every vector you use when calculating a sum of vectors. That is, you must measure the length of each vector with a ruler and use a protractor to measure the angle associated with each vector. You say there are several angles for each vector? You are correct. Use a convention for indicating direction that you are already familiar with in order to clear up the ambiguity. Use north as the direction vertically upward and south for the direction vertically downward. East will be the direction to the right of the page and west the direction to the left. What if the direction angle is between two of these? Measure the acute angle between the vector and a horizontal line you draw through the foot of the vector.

Suppose the angle is 30° upward from the horizontal. The direction will be either East30°North or West30°North. That is, the direction is upward and to the right of the page or upward to the left of the page.

You can list the direction as E30°N. That means you stand at the foot of the vector and face east, and turn 30° toward north and you are facing the direction the arrow is pointing. Similarly, W30°N means, face west and turn 30° toward north and you are facing in the direction the arrow is pointing. This method of indicating the direction of a vector is the convention used early in the study of vectors. This way is used probably because using maps of the countryside is part of almost everyone's experience. There are other conventions for representing direction such as the heading of a ship or aircraft, which are checked constantly by the pilot or captain to make sure they wind up in the right place.

Physics Phun

I measured the vectors in Figure 3.5 using a ruler and a protractor. Here is a record of those measurements: $\vec{A}$ = 1.9 cm E, $\vec{B}$ = 1.1 cm W55°N, and $\vec{C}$ = 2.4 cm E67°S. Notice that both the magnitude and direction must be given to specify a vector. Use those same instruments and make the same measurements to see if you agree with measurements I recorded. Then add these vectors together using the rules stated in this section along with the aid of the example for adding three vectors. Hmm, 1.9 + 1.1 + 2.4 no longer equals 5.4 in this new world of vectors. Strange things are happening.

A New Subtraction Also

If you thought that adding vectors was weird, you have a surprise coming. The reason vectors are so strange is because we are used to using scalars. Changing to a new idea is difficult. Any kind of change is uncomfortable at times and as human beings we naturally resist change. When you think of addition, you think of scalar addition unless you are reminded that you are working with vectors. When you think of subtraction, you think of the "seven take away three equals four" kind of operation. As you can imagine, subtracting vectors is not as easy as that. Neither was it easy to think of anything but "three plus three equals six" kind of addition. I think that you found for vectors that three plus three is not necessarily equal to six! There is one good thing about subtraction, though, and that is you do find the difference in two vectors. As it turns out, you must always work with only two vectors at a time when you are carrying out the operation of subtraction.

Suppose you want to find the difference in two vectors, $\vec{A}$ and $\vec{B}$. The operation of subtraction can be done in one of two ways: either $\vec{A} - \vec{B}$ or $\vec{B} - \vec{A}$. The rules for subtraction are (1) write down which difference you want to calculate, (2) construct the vectors so that they have their feet together in the same plane—that is, their feet are at the same point—and (3) draw a vector from the head of the second to the head of the first. The notation used in step one establishes identification of the *first* vector and the *second* vector. The first one listed in that notation is the first, and the *second* is the second vector listed in the notation. That is, in the notation $\vec{A} - \vec{B}$, $\vec{A}$ is the first and $\vec{B}$ is the second. That means the difference vector will be drawn from the head of $\vec{B}$ to the head of $\vec{A}$ in that order. It sounds complicated, but after an example and a bit of practice you will see that it is not difficult.

Subtraction may seem a little nutty but you will use this operation to develop valuable information soon. Look at Figure 3.6, where you are given two vectors and an example of subtraction worked out for you. As a check, you may look to see if vector $\vec{B}$ plus the new vector $\vec{A} - \vec{B}$ would add to the original vector $\vec{A}$.

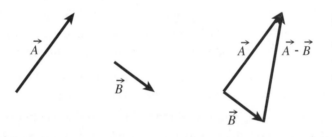

Figure 3.6

Subtracting vectors is an illustration of the rules for subtraction.

Physics Phun

Use the vectors in Figure 3.6 to construct the vector difference $\vec{B} - \vec{A}$. Write your answer in the form $\vec{B} - \vec{A} = ?$. Remember to give both magnitude and direction for your answer.

Complications of Multiplication in Vector Algebra

So far we have constructed some important parts of vector algebra. Up to this point you know (1) the definition of a vector, (2) what equal vectors are, (3) what vector addition is, and (4) how to subtract vectors. The review of one more operation, vector multiplication, will provide us with all the vector algebra we will need to pursue the development of many ideas in physics.

There are several types of vector multiplication but we will consider only *scalar multiplication*, which is the process of multiplying a vector by a scalar. Chances are you remember when you first learned to multiply. It was probably introduced as a quick way to add. That is, 4×3 is just a quick way of writing 3+3+3+3. You get the same answer both ways but 4×3 is faster and less complicated to write.

Suppose you are given a vector $\vec{A}$ and you are asked to calculate $4\vec{A}$. You probably think, knowing what you know at this point, that would be $\vec{A}+\vec{A}+\vec{A}+\vec{A}$. You already know how to add vectors and you know that scalar multiplication is probably just a quick way to add. You are correct. The vector $4\vec{A}$ is a vector four times as long as $\vec{A}$ and has the same direction as $\vec{A}$. It is important to note that scalar multiplication produces a vector and not a scalar. In general, for any scalar, say n, $n\vec{A}$ is a vector having the same direction as $\vec{A}$ but n times as long as $\vec{A}$. That means that $\frac{1}{4}\vec{A}$ is a vector with the same direction as $\vec{A}$ but only $\frac{1}{4}$ the length of $\vec{A}$. Did you notice that multiplying by $\frac{1}{4}$ is the same as dividing by four? So you can divide a vector by a scalar and produce a vector so long as the scalar is not 0.

Plain English

Scalar multiplication is the product of a scalar and a vector that results in a vector with a magnitude determined by the scalar. The direction of the product is the same as the original vector if the scalar is positive and opposite the direction of the original vector if the scalar is negative.

Another exciting observation is this: suppose $n = -1$. Do you see that the vector $\vec{A}$ still has the same magnitude? What has changed? Right! Only the direction has changed, so $(-1)\vec{A}$ is a vector with the same length as $\vec{A}$ but a direction that is opposite that of $\vec{A}$. You now have a new vector $-\vec{A}$ that is just the opposite of $\vec{A}$ and behaves just like scalar opposites. That is, –3 is the opposite of 3 and 3+(–3)=0. So (–)+=0. You have discovered a strange new vector, 0, that is a vector with no direction and 0 length.

If you enjoy algebra (and some of us weirdos do, you know), then vector algebra is becoming more interesting all the time. We can expand the algebra now to include scalar multiplication, legal scalar division, the negative of a vector, and also a 0 vector. You can check all of this nonsense if you like. Now that you know what the negative of a vector is, you can define the negative of either vector in Figure 3.6 and add it to the other vector. If our definition of scalar multiplication is correct, adding the negative should yield the same result as the complicated rules stated for subtraction. You want to remember those rules even though you have a check using the negative of a vector. The rules for subtraction will serve to save you time and provide you with a special tool for pursuing a review of any quantity that might have the properties of a vector.

Is Breaking Up Hard to Do?

You know that you can add as many vectors together as you wish. All you have to do is to preserve magnitude and direction and lay out one vector in a plane. Then you join the foot of the next vector to the head of the previous one until you have completed addition of all the terms. Finally, you join the foot of the very first vector to the head of the last vector, in that order, and thus identify the resultant. That process might involve an infinite number of parts to determine the resultant.

Consider the possibility of an inverse process. Suppose you start with a vector, a resultant if you want to call it that, and you are confronted with the problem of finding the parts that were combined to make up the given vector. You could say, "Okay, I'll play your silly game," then proceed to draw vectors foot to head beginning at the foot of the original vector and ending at the head of the original vector. You would be correct, because there are an infinite number of ways to solve the problem.

The process of identifying the parts of a given vector is called *resolution*. The parts are called the *components* of the vector. This might be the same as asking for change for a dollar. There are many possible combinations that might add up to a dollar. But suppose that I want to use the change in a vending machine. Then pennies would not be very helpful. In the same way, there are some vector components that might prove to be more helpful than others. While your first solution was a good one and not too difficult, looking for and naming specific components can be a challenge. As we proceed in our review of physics, you will find that resolving a vector into certain components can be very helpful in solving problems. The ability to identify certain components can provide you with another tool for clear critical thinking.

Plain English

Resolution is the name of the process of identifying the parts of a vector. The **components** of a vector are those parts whose sum is the given vector.

Two components you want to be prepared to find are the two that are perpendicular to each other. These are easy to locate if you remember from geometry that the projection of a point on a plane is the foot of a perpendicular drawn from the point to the plane. So if you want to find the projection of a line segment on a plane, drop a perpendicular from the endpoints of the line segment to the plane and connect the two points in the plane with a line segment. I have constructed that bit of geometry for you in Figure 3.7. Notice that the line segment is labeled *AB* and its projection on *l*, a line in a plane perpendicular to the page is labeled *A'B'*.

Figure 3.7

The projection of AB on l illustrates the projection of a line segment on a plane.

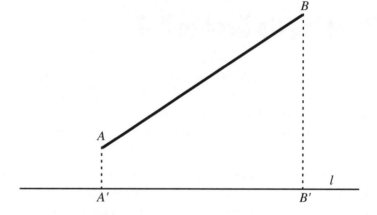

The information summarized in Figure 3.7 can be used to find two mutually perpendicular components of any vector. Given a vector $\vec{A}$, construct two mutually perpendicular lines through the foot of $\vec{A}$. Construct the projection of the vector on each of those lines. Place an arrowhead on each of the components at the points on *l* corresponding to the projections of the tip of the arrow representing $\vec{A}$.

You have constructed two components of $\vec{A}$ that are mutually perpendicular. Among many other things, these components will enable you to calculate the magnitude of $\vec{A}$ by using the Pythagorean Theorem. Is that another dirty word? It means that the magnitude of $\vec{A}$ is the square root of the sum of squares of the magnitudes of the components. Figure 3.8 will give you a detailed look at the resolution of $\vec{A}$ into two components that are mutually perpendicular.

Figure 3.8

The resolution of $\vec{A}$ into two mutually perpendicular components illustrates the identification of two crucial vector components.

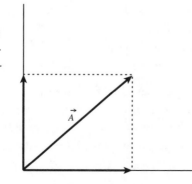

Position and Displacement

A study of motion makes you aware that an object moves in relation to some point or place. Describing motion means that you assume a certain point or place, often referred to as a frame of reference, and you tell where an object is with respect to that place. A point on the earth's surface is a good reference point that you are familiar with, so use that frame of reference unless stated otherwise. Assume we are using a small enough part of the earth so we may act as if it were a flat plane.

The study of vectors enables us to discuss positions and motions in a plane where earlier we discussed motion along a straight line. As you can see, vectors will enable you to discuss the familiar motion along a straight line as well as motion in a plane perpendicular to the surface of the earth, parallel to the surface of the earth, and in fact in any plane between. Vectors free us from the straight-line world, enabling us to tackle the universe. Well, the universe near the surface of the earth at least. We won't go into orbit yet but that is coming!

Remember the couch potato? You can begin applying to him this fresh approach of describing motion. Let his initial *position* be your reference point. You can say his initial position, a vector quantity, is $\vec{X}_0$ and his final position is $\vec{X}_f$. When he moved from one end of the couch to the other, he had a change in position of $\vec{\Delta X} = \vec{X}_f - \vec{X}_0$.

Notice that this is the difference in two vectors and that means that $\vec{\Delta X}$ is a vector drawn from the head of position $\vec{X}_0$ to the head of position $\vec{X}_f$. Earlier we said that the couch potato experienced a change in position, which is obvious when we use the vector nature of position. We called Δx the distance traveled along the couch, and that is correct because Δx is a scalar quantity. $\vec{\Delta X}$ is a vector quantity, which has magnitude, called distance traveled, and direction. In physics, $\vec{\Delta X}$ is the vector called *displacement*. You might note that the magnitude of position is also distance but not distance traveled.

> **Plain English**
>
> **Position** is a vector used to locate a point from a reference point or frame of reference. **Displacement** is a vector and is defined as a change in position.

Understanding Velocity

Back to that couch potato again. You know that I used that simple situation to define average speed. In the same frame of mind, and using the same frame of reference as we did before (even though we did not call it that), I will define a new notion. This idea is similar to average speed, but contrary to experience with motion in the past, it is very different. Take the displacement of the couch potato and divide by the time required to make the displacement and you will see what I mean.

$\frac{\overrightarrow{\Delta X}}{\Delta t} = \overrightarrow{V}$ is the equation resulting from that operation, and in words it states that the displacement divided by the time is equal to the average *velocity*. Velocity is a vector quantity, as you can see. It results from the division of a vector $\overrightarrow{\Delta X}$ by the scalar Δt or, in this case, t the time for the displacement to take place. The velocity vector has the same direction as $\overrightarrow{\Delta X}$ and has the speed for a magnitude.

Velocity is a vector quantity while speed is a scalar quantity. Velocity and speed are not the same things, and they are not interchangeable in general. Occasionally a discussion might involve objects in motion along a straight line in one direction; then speed and velocity will have the same values. The instant the object changes direction, only velocity is defined. Velocity only can be used when an object moves off the straight line into a plane or changes direction and moves backwards. For example, you might say that your car has a speed of $60\frac{mi}{hr}$, the reading from your speedometer, which is perfectly all right. If you say that your car is traveling $60\frac{mi}{hr}N$, then it cannot be read from your speedometer, and it is not speed. The velocity of your car in that case is $60\frac{mi}{hr}N$. The velocity reading would be a combination of a reading from the speedometer and a reading from a compass. The important thing to remember is that speed and velocity are two different animals. You will get plenty of practice with these ideas in later sections.

Plain English

Velocity is the rate of change of displacement. Velocity is a vector quantity that has the same direction as the displacement, and its magnitude is the speed.

The Least You Need to Know

◆ Distinguish scalars from vectors.

◆ Use vector algebra to discuss vector quantities.

◆ Resolve a vector into mutually perpendicular components.

◆ Discuss motion in terms of position and displacement.

◆ Recognize and use the vector nature of velocity.

Chapter 4

Motion in Two Dimensions

In This Chapter

- ◆ Walk around until you're "displaced"
- ◆ The velocity of planes and boats
- ◆ Acceleration with a change in velocity
- ◆ Speed, velocity, and centripetal acceleration

Are the new terms introduced in the last chapter labels of ideas that will help us to gain understanding of concepts in physics? That is, do displacement and velocity have real meaning or are they just fancy names for show? In this chapter I will answer these questions, with your help, as we consider several examples. Questions like these should be addressed because they often lead to tests in the laboratory. However, some examples we will consider involve situations that you do not want to experiment with. The velocities may be impractical or even dangerous if actually tried. Often special situations and conditions are set to illustrate a point. Don't try one of these at home unless it is obvious to you that it will not endanger your health.

Understanding Displacement

An example of a real application of displacement is taking a walk. Suppose you take the following route around the neighborhood to enjoy the out-of-doors. Start at your front doorstep, walk 3.0 blocks north, then 6.0 blocks east, then 8.0 blocks south, 2.0 blocks west and finally 1.0 block north. You have walked 20.0 blocks, a considerable distance. Calculate your displacement from your front doorstep to the spot marking the end of your walk.

Newton's Figs ───────

Preserve the magnitude and direction of all vectors when you perform any type of operation with vectors.

You know that displacement is a vector quantity, so the individual displacements must be added together to find the displacement from your front door. Choose a convenient scale for constructing the magnitudes of displacements, such as 1 cm represents 1 block. Use the convention introduced in the last chapter to record directions. Construct your own solution and compare your results with my solution given in Figure 4.1.

Figure 4.1

The solution to a displacement problem shows how different distance traveled is to displacement.

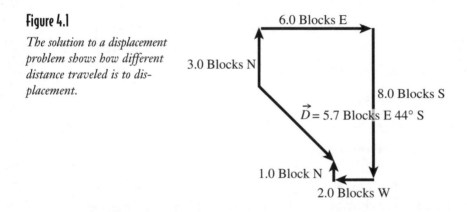

6.0 Blocks E

3.0 Blocks N

8.0 Blocks S

$\vec{D}$ = 5.7 Blocks E 44° S

1.0 Block N

2.0 Blocks W

Recall that because of your data your answer will have two significant figures as well as two parts, magnitude and direction. The technique outlined here requires the use of a ruler and a protractor, and your solution involves a scale drawing. This technique is called a *graphic solution*.

I found a displacement of 5.7 blocks E44°S. When you measure the magnitude with the ruler, you find 5.7 cm, where 1 cm represents 1 block on your scale drawing. Drawing a horizontal line through the foot of the displacement vector allows the measurement of the angle with a protractor. Another way of solving the problem is to add all north and south displacements together and then add all east and west displacements together. The result obtained turns out to be two mutually perpendicular components of the displacement vector.

The magnitude of the displacement vector can be calculated by using the *Pythagorean Theorem*, where the magnitudes of the components make up the legs of a right triangle. The same right triangle will enable you to measure the appropriate angle with a protractor. When I used this method, I found 4.0 blocks south and 4.0 blocks east for the two components. I leave this last solution for you to think about.

Plain English _____

A **graphic solution** is achieved by using drawing instruments to construct a scale drawing. A ruler provides the measurement of magnitude to scale and the protractor enables you to measure angles. The **Pythagorean Theorem** is a theorem from plane geometry relating the legs and hypotenuse of a right triangle. The square of the hypotenuse is equal to the sum of the squares of the two legs. The legs are the mutually perpendicular sides and the hypotenuse is the side opposite the right angle in the right triangle.

If you can solve this problem using the two methods suggested here, you will not only have a check on your work but you have a head start on similar vector problems discussed later. You can see that displacement is not the distance traveled. It has magnitude and direction resulting from the addition of all of the displacements on your trip. Compare the magnitude of displacement with the distance traveled, 20 blocks!

Understanding the Velocity of a Plane

You have probably noticed from time to time that a small plane flying overhead has its nose pointed in one direction while the plane is moving in an entirely different direction. Ever wonder why? You can figure that out with your knowledge of velocity. Enjoy the following example: What is the velocity of a plane relative to the air if it is to fly due north at 320.0 km/hr relative to the ground while the air is moving east at 80.0 km/hr relative to the ground? You know that the plane is moving through the air and that the air is moving above the ground, taking the plane with it. If the plane is to fly in a straight line above the earth, it must point its nose in a direction that will balance the effects of the motion of the air. The direction that the plane points its nose is called the "heading" of the aircraft. The solution to this problem will be the heading and the air speed of the plane. The solution is found in Figure 4.2 by using a graphic solution.

The heading indicates that the nose of the plane will always be pointed in one direction for the duration of the flight while the plane moves in a straight line above the earth in a different direction. That is, in my solution, the nose will always point N15°W and the plane moves through the air at 330 km/hr. Notice that the plane's velocity relative to the air has a westward component which is the negative of the velocity of the air relative to the ground. This component nullifies the effects of the movement of the air. The northern component of the plane's velocity relative to the air enables the plane to fly straight as an arrow! The vector equation used is $\vec{V}_{pg} + (-\vec{V}_{ag}) = \vec{V}_{pa}$. In words, the velocity of the plane relative to the ground plus the opposite of the velocity of the air relative to the ground equals the velocity of the plane relative to the air. See how efficient the vector algebra proves to be?

Figure 4.2

The velocity of an airplane as determined by component velocities.

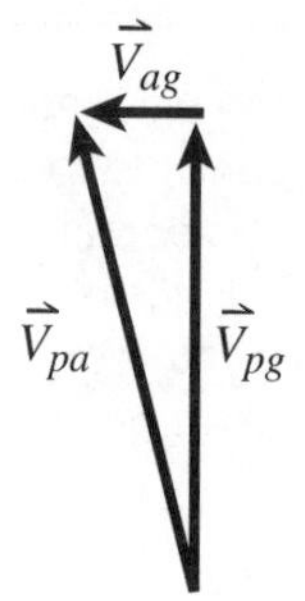

Suppose the flight takes 1.00 hr. Calculate the displacement of the plane relative to the air. All of the motion is uniform so we can describe the uniform motion as we did in an earlier section, but this time for vector quantities. Using the vector equation for velocity, $\vec{V}_{pg} + (-\vec{V}_{ag}) = \vec{V}_{pa}$, multiply both sides of the equation by the scalar t for time and find $\vec{V}_{pg}t + (-\vec{V}_{ag})t = \vec{V}_{pa}t$. Using the relationship developed earlier, this equation becomes $\Delta\vec{X}_{pg} + (-\Delta\vec{X}_{ag}) = \Delta\vec{X}_{pa}$ or $\vec{D}_{pg} + (-\vec{D}_{ag}) = \vec{D}_{pa}$. Both of these equations state in words the displacement of the plane relative to the ground plus the opposite of the displacement of the air relative to the ground equals the displacement of the plane relative to the air. Using the results of our solution in Figure 4.2 and the additional information that the time is one hour, the solution is

$$\vec{D}_{pa} = \vec{V}_{pa}t = 328km \, / \, hrN15°W \times 1.00hr = 328kmN15°W.$$

You can also calculate each displacement and then add the vectors graphically and find that you get the same result. I will continue to demonstrate different ways of solving problems so that you will have many options of models to use to construct your own solution. You may also choose to solve problems at least two different ways in order to check your results.

Understanding the Velocity of a Boat

Suppose you can row a boat in still water at 4.0 mi/hr and you want to travel across a river that is flowing from north to south at 5.0 mi/hr. If you maintain your velocity directly across the river for 10.0 minutes, you reach the other side. What is the velocity of the boat relative to the ground? How far down the river will the boat land? What is the minimum boat velocity (relative to the water) required to travel in a straight line across the river? As usual, analyze the problem before attempting a solution. There are several questions to be answered but all of them involve uniform motion. Approach the questions one at a time. List the information given in terms of the type of motion and the type of quantities involved.

The boat will move through the water at 4.0 mi/hr whether the water is moving or not: that is, $\vec{V}_{bw}$ = 4.0 mi/hr E assuming you start on the western side. The river has a uniform velocity relative to the ground of $\vec{V}_{wg}$ = 5.0 mi/hr S. The time can be tricky to interpret unless you note that we should express it in hours so that all information is in hours and miles. The time is $t = 10.0\,\text{min} \times \dfrac{1hr}{60\,\text{min}} = 0.17hr$.

Notice the use of a defining equation to write minutes in terms of hours. We want to find $\vec{V}_{bg}$, $\vec{D}_{wg}$ for time t, $\vec{V}_{bw}$ for the boat to travel east in a straight line.

Refer to the vector diagram in Figure 4.3 in order to visualize the problem and begin a solution. The diagram is a graphic solution where I assigned 1 cm to represent 1.0 mi/hr.

Newton's Figs

If you did not interpret the problem in this section as described, don't worry, you will develop this type of critical thinking and focused reading by the time you complete this book. Remember, through the book I am your private tutor.

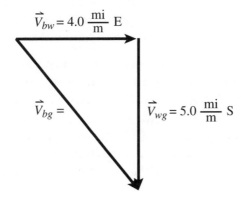

Figure 4.3

The velocity of a boat crossing a stream as an example of component velocity solution.

$\vec{V}_{bw}$ = 4.0 $\dfrac{\text{mi}}{\text{m}}$ E

$\vec{V}_{bg}$ =

$\vec{V}_{wg}$ = 5.0 $\dfrac{\text{mi}}{\text{m}}$ S

Use any scale you choose to solve the problem. Note that we do not need a scale drawing to find $\vec{D}_{wg}$ because we know the time and velocity. Therefore, $\vec{D}_{wg} = \vec{V}_{wg}$ t=(5.0 mi/hr S)(0.17 hr)=0.85 mi S. Does it surprise you that the boat lands 0.85 mi downstream from where it started? It is on the other side of the river of course. We answered one of the questions. There is no reason to follow any particular order to answer all the questions.

The diagram shows that $\vec{V}_{bg} = \vec{V}_{bw} + \vec{V}_{wg}$ and the graphic solution is 6.4 mi/hr *S39°E*. The diagram also shows that $\vec{V}_{bg} + \left(-\vec{V}_{wg}\right) = \vec{V}_{bw}$ where $\vec{V}_{bg} = 4.0mi\,/\,hrE$ for the case where the boat is to travel east in a straight line. You can construct a graphic solution or you can see by the symmetry of the problem $\vec{V}_{bw} = 6.4mi\,/\,hrN39°\,E$. That means that if you want to go straight across the river you must paddle like a stern-wheeler or get a motor in order to maintain the required speed through the water.

You probably noticed the airplane problem from the previous section and the boat problem here are very similar. Both involve an object moving through a medium while the medium is moving relative to the ground. You need to specify the frame of reference to clarify your solution to a motion problem. Since all motion is relative, you see that it is essential to specify relative to what point or to what frame of reference. When you did that in the last solution, you arrived at an understandable solution by stating motion relative to the water or to the ground.

Understanding Acceleration

The situation I set up here for your study is not practical or maybe not even possible. Remember the speed demon in the Dodge Dart that can do sixty in a quarter? If you drive very much you have probably seen that guy on the road. Furthermore, you give him all the road he wants. Consider a situation where he was driving toward an intersection at 20.0 km/hr E when he made a sharp turn at the intersection without changing speed and is now traveling at 20.0 km/hr N. Quite a stunt, don't you think?

I warned you not to try some of these ideas at home or on your own. This is one of those examples that could be dangerous for your health. If this could be done, can you explain the tremendous change you would observe? The speed remains the same but it is clear that the velocities are not the same. The velocity must have changed. If the velocity changed, what does that imply? In Figure 4.4, you see a vector diagram of the final velocity and the initial velocity along with the calculated change in velocity. The change in velocity took a small amount of time *t*, or Δt if you prefer.

Figure 4.4

The calculated change in velocity implies acceleration.

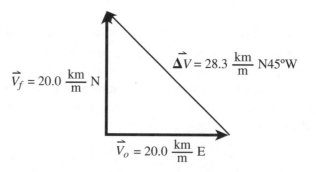

The change in velocity $\overrightarrow{\Delta V}$ that took place in time Δt=3.00 s, for instance, implies that there is acceleration $\overrightarrow{a} = \dfrac{\overrightarrow{\Delta V}}{\Delta t}$. You calculate that quantity from the diagram and along with the hypothetical time see that there definitely is vector acceleration $\overrightarrow{a} = \dfrac{28.3 km\,/\,hr N 45° W}{3.00 s} \times \dfrac{1hr}{60\min} \times \dfrac{1\min}{60 s} \times \dfrac{10^3 m}{km} = 2.62\,\dfrac{m}{s^2}\,N 45° W.$ The exciting thing to realize here is that we have an object with a constant speed and yet it accelerates! You know that a direction change of 90° in 3.00 s is very drastic. I used these figures to make it easy to calculate acceleration. The change in velocity of this example would probably take a fraction of a second. If you have the stomach for it, you can find a ride at an amusement park that will approximate this example. You are alert to the fact that acceleration can exist without a change in speed. A more realistic example of this is presented in the next section.

Uniform Circular Motion

In an earlier section we defined uniform motion as characterized by a constant velocity. The focus was on straight-line motion to begin with. In a previous section, I changed the focus to uniform motion in a plane. The hypothetical constant speed of the speed demon may not have been very convincing, but it still illustrates how to describe uniform motion in a plane. You will find that the discussion you are exposed to here will be much more convincing because it involves real physical situations.

At some time in your experience you have whirled some object about your head at a constant speed. A string tied to the object and the other end of the string held in your hand defined the circular path of the object. You probably even released the object at some point and observed it to fly away in a straight line that is tangent to the circular path at the instant of release. In case that was a long time ago or maybe you haven't experienced that yet, you will want to keep all of the details in mind. Since a circle is a plane figure and uniform motion in a circular path is motion in a plane, you use vectors to describe the motion.

Speed and Velocity in a Circular Path

I ask you to revisit these ideas once again in another context. The distinction between the speed and velocity will reveal to you a most amazing notion in physics. There is one and only one vector acceleration that results from the motion of the object in a circular path at a constant speed. The speed can be calculated in terms of the time required for the object to make one *complete revolution* or *circular path*.

Given that the speed is constant, the time required to make one complete revolution is called the *period of motion*.

Plain English

One **complete revolution** or **circular path** is the circumference of the circle. The **period of motion** is the time required to complete one revolution.

Since the period is a very special time in physics, it is usually represented as T in seconds or some other unit of time. The speed then would be the distance traveled in one revolution divided by the time to complete one revolution. The geometry of the circle allows you to calculate the distance traveled in one revolution by calculating the circumference of the circle. The speed is calculated to be $v = \dfrac{2\pi R}{T}$, where R is the radius of the circle and T is the period of the motion.

Have you noticed that mud on a spinning tire flies off tangent to the tire at the instant it leaves the tire and travels in a straight line? That is, the straight-line path is perpendicular to a radius of the tire. That means a 90° angle is formed between the radius of the tire and the straight-line path of the mud. Whenever any object is moving in a circular path at a constant speed, it has a velocity that is tangent to the circular path at every instant.

I use $\vec{V}$ to represent velocity and v to represent speed. A position vector $\vec{R}$, length R, that is drawn from the center of the circle along a radius to a point on the circle indicates the position of the object on the circular path relative to the center of the circle. Two positions of an object moving in a circular path at a constant speed have been identified in Figure 4.5. The two positions can be anywhere on the path.

I chose $\vec{R_0}$ to be the position at the beginning of a time interval Δt or t, either may be used to represent a small time interval. $\vec{R_f}$ is the position of the object at the end of the time interval. The corresponding velocities of the object are represented by vectors tangent to the path at those positions. All of this information is used along with Figure 4.5 to guide you through calculations of quantities implied by the physics of this situation.

Centripetal Acceleration

Suppose you are given that the speed of an object is v. That means that $v = v_0 = v_f$ because the speed is constant. Look at the diagram in Figure 4.5. Notice that d is a chord and s is a corresponding arc of the circular path. Recall from geometry that for small angles the length of d is approximately equal to the length of s and for that reason you feel free to substitute s for d when the need arises.

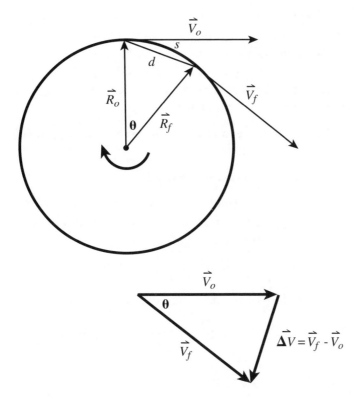

Figure 4.5

The object with uniform circular motion is located by position vectors $\vec{R}_0$ and $\vec{R}_f$.

The angles θ and ϕ are small angles, as small as you want to make them. Hence, approximating the length of the chord with the length of the arc is legitimate. I will refer to the angles in terms of their right sides and left sides. It is clear what that means: If you imagine standing at the vertex of an angle and looking out along its sides, the side on your right is the right side and the one on your left is the left side. Feel free to refer to all of this information as I develop the algebraic expressions from the geometry of the figure and the physics of the situation.

You can see that even though the speed is constant the velocity changes at every instant. The change in velocity can be calculated as was done before. That is, construct the velocity vectors with their feet together so that you can find $\vec{V}_f - \vec{V}_0 = \vec{\Delta V}$. Remember that the magnitude of $\vec{V}_f$ equals the magnitude of $\vec{V}_0$ equals v but the magnitude of $\vec{\Delta V}$ is ΔV, which the length of the change in velocity vector determined by the geometry of the triangle in the construction. It is understood that it is not Δv because the speed v is constant. It does not change. Since there is a change in velocity, there must be acceleration. You are about to calculate that acceleration by using all of the information cited.

There are two triangles of interest here. One is a distance isosceles triangle with sides R, R, d and vertex angle θ. The other is a velocity isosceles triangle with sides v, v, ΔV and vertex angle ϕ. In the diagram of the circular path of Figure 4.5, the position vectors are perpendicular to the velocity vectors so the right side of θ is perpendicular to the right side of ϕ and their left sides are perpendicular as well. Each algebraic statement is followed by justification for the statement:

$\theta = \phi$ Two angles are equal in measure if they have their sides perpendicular right side to right side and left side to left side.

θ and ϕ Are vertex angles of two isosceles triangles.

Therefore, the distance triangle is similar to the velocity triangle. The reason is because two isosceles triangles are similar if the vertex angle of one is equal in measure to the vertex angle of the other.

$\dfrac{\Delta V}{d} = \dfrac{v}{R}$ Corresponding parts of similar triangles are proportional.

$\Delta V = \dfrac{dv}{R}$ Multiply both sides of the previous equation by d.

$\dfrac{\Delta V}{\Delta t} = \left(\dfrac{d}{\Delta t}\right)\left(\dfrac{v}{R}\right)$ Divide both sides of the previous equation by Δt and group the factors.

$\dfrac{\Delta V}{\Delta t} = \left(\dfrac{s}{\Delta t}\right)\left(\dfrac{v}{R}\right)$ Substitute s for d.

$\dfrac{\Delta V}{\Delta t} = (v)\left(\dfrac{v}{R}\right)$ The speed of the object is $\dfrac{s}{\Delta t}$, the distance traveled along the arc divided by the time.

$\dfrac{\Delta V}{\Delta t} = \dfrac{v^2}{R}$ and $a = \dfrac{v^2}{R}$ Substitute the definition of acceleration.

Remember that all of the algebraic statements utilize scalar quantities. The acceleration is a vector quantity. It has the same direction as the change in velocity vector. We have just calculated the magnitude of this acceleration to be equal to the square of the speed divided by the radius of the circle. You can see that the change in velocity vector becomes very close to being perpendicular to the velocity vector as the angles θ and ϕ get smaller and smaller. Therefore, the direction of the change in velocity vector, and the acceleration, would then be directly opposite the direction of the position vector, always pointing toward the center of the circular path.

For these reasons, the acceleration for uniform circular motion is called *centripetal acceleration* because it is directed radially inward toward the center of the circle.

Centripetal acceleration means center-seeking acceleration. It is the one and only acceleration that results in uniform circular motion. If the acceleration is always directed perpendicular to the instantaneous velocity vector, then the motion will end up being a circle, and the acceleration will point towards the center of that circle. Remember that, because it will enable you to explode a myth that you probably grew up believing.

Plain English

Centripetal acceleration is the center-seeking acceleration of a body with uniform circular motion. Furthermore, it is the only acceleration defined by uniform circular motion.

Any time you are discussing centripetal acceleration it is understood that the direction is radially inward toward the center of the circle. At this point we will express centripetal acceleration completely by $\vec{a}_c = \dfrac{v^2}{R}$ toward the center of the circle. If you are not given the speed, you can express speed in terms of period and find $\vec{a}_c = \dfrac{\left(\dfrac{2\pi R}{T}\right)^2}{R} = \dfrac{4\pi^2 R}{T^2}$ toward the center by substituting $\dfrac{2\pi R}{T}$ for v in the expression $\vec{a}_c = \dfrac{v^2}{R}$ toward the center. Since the direction of centripetal acceleration is toward the center of the circular path, you will probably be concerned only with its magnitude given by the two expressions $a_c = \dfrac{v^2}{R}$ and $a_c = \dfrac{4\pi^2 R}{T^2}$. Check both expressions for the correct units for acceleration in all three systems of measurement.

Physics Phun

1. Calculate the speed of an object moving in a circular path of radius 2.0 m if it has centripetal acceleration of $2.6\dfrac{m}{s^2}$.

2. Calculate the period of the motion of the object in (1).

3. Calculate the radius of the circular path of an object if it has centripetal acceleration of $2.0\dfrac{m}{s^2}$ and a speed of $3.0\dfrac{m}{s}$.

The Least You Need to Know

◆ Construct a sum of displacements.

◆ Find the difference in two velocities.

◆ Solve velocity and displacement problems.

◆ Calculate and interpret centripetal acceleration.

Part 2

When Does Something Move?

You may be thinking that something moves when you step on its tail. Something moves when you swing at it with a fly swatter. Specifically, a baseball moves when you hit it with a bat. That is the type of movement that we are interested in here. In this part, you look at the causes of motions of very small objects and very large objects near the surface of the earth. Also the motions of objects falling to the earth and motions of objects falling around the earth are considered. The motion of an object falling around the earth is used to determine the mass of the earth. All of these ideas are yours for the taking.

Pushes and Pulls

In This Chapter

- ◆ Helpful guides to discovery
- ◆ A practical application of a vector
- ◆ Useful parts
- ◆ Old techniques for handling force

So far the focus has been on the description of motion. In textbooks, you find kinematics includes all the topics treated thus far in this book. Kinematics is the study of the description of motion. The next step is the exploration of the causes of motion, known as dynamics.

Everybody knows that if you want to make an object move, you either push on the object or pull the object. The push or pull you apply to the object is called a force. You know that if you push the object it tends to move away from you and if you pull it tends to move toward you. That means that force is a vector quantity. Force is a derived quantity and that means you will become familiar with the units of force in each of the systems of measurement. You also know that if the object moves, it can move in one of two ways that we are now able to describe. It will have either uniform motion or uniformly accelerated motion. Addressing all of these ideas about force is quite a task. It is helpful to begin with some tools that will give you insight into the nature of force.

Mathematical Models and Physical Relationships

I found that no matter how many years of mathematics my physics students had taken, most had not made the connection between algebra and their studies in physics. I assumed the responsibility to make the connection for them. In order to gain as much as you can from ideas that I will share with you, it will be most helpful for you to review ideas involving graphical analysis in an algebra book.

There are many references to the different relationships mentioned throughout this book. I think you will agree that the time you spend mastering these ideas is time well spent. You need to have this background at this time so that you will be comfortable with my approach to the development of the ideas of force. You participate in the process rather than waiting for me to tell you what happens next. You will know when something makes sense if you have sufficient evidence to make a decision about it. You will know what evidence to look for and you will decide!

There are many mathematical models used by scientists in general and physicists in particular. I choose to use three relatively simple ones to help you begin to get inside the head of the physicist. Imagine how scary that can be! These models will provide you with creative tools that help you to master most of the material in your physics courses.

Experimental data are analyzed using families of graphs to discover a relationship, if one exists, between the variables. The experimental data are expressed in terms of two variables. The experimenter controls one variable and the other variable responds to reveal a relationship, if one exists, between the two variables. So that is the plan followed here:

♦ Linear Relation: $y=mx$, where m is the slope of the line.

♦ Power Relation: $y=Bx^n$.

♦ Exponential Relation: $y=BA^x$.

In all of the equations for the models, y is a dependent variable and x is an independent variable, while A, B, and m are all constants. Associated with each of the equations is a graph or several graphs depending on the values assigned to the variables.

Remember that these are algebraic models. The physicist often borrows a mathematical model to explain a scientific principle or law. When that happens, variables that fit the experiment enable a student to understand the mathematical statement of the scientific relationship. Often the scientific statement is patterned after the mathematical statement but will have values of variables that do not precisely satisfy the equation for the mathematical model. I call the variables used by scientists the reacting variable instead of the dependent variable and managed variable instead of independent variable so that the model, the ideal, will not be confused with the scientific statement.

Applying Graphical Techniques to Experimental Data

You know that a force is a push or pull, it is a vector quantity, and it is a derived quantity. You know that if you apply a force to an object it will accelerate. Recall the speed demon leaving the parking lot like a flash. I did an experiment using five arbitrary unit forces and five arbitrary unit masses and measured the acceleration under circumstances given in the following table. The acceleration is the reacting variable and the force is the managed variable.

Acceleration Versus Force (mass is held constant of 1 unit)

Acceleration (m/s²)	Force (Arbitrary Unit)
0.20	1.0
0.40	2.0
0.60	3.0
0.80	4.0
1.0	5.0

The graph of the data in Figure 5.1 looks like a straight line that would pass through the origin. Using the ideas just reviewed, it is logical to suggest that $A \propto f$ when the mass of the object is held constant. That means that the acceleration increases as the force applied to the object increases. That information is sufficient at this stage because there is a third variable, m, involved.

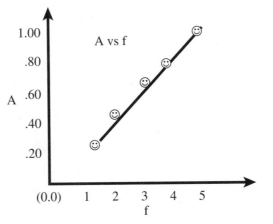

Figure 5.1

The graph of A vs. f suggests a direct variation.

The next part of the experiment was done by making the mass, m, the managed variable and the acceleration, A, the reacting variable. The acceleration is again measured in m/s² and the mass in arbitrary units. The table is a record of the data.

Acceleration Versus Mass (force held constant of 1 unit)

Acceleration (m/s²)	Mass (arbitrary units)
0.800	1.0
0.400	2.0
0.270	3.0
0.200	4.0
0.160	5.0

The force applied is one unit in each trial. The graph in Figure 5.2 looks like a power relation where the exponent is a negative integer. Graphing techniques are used to check that possibility by beginning with $n = -1$. The graph in Figure 5.2 suggests that there is an inverse variation so the logical place to begin the search for the correct power is with the case $n = -1$. That is, it appears that $A \propto \dfrac{1}{m}$ and graphing this information will help you to decide whether that is the relationship you are looking for.

Figure 5.2

The graph of acceleration versus mass with a constant force of one unit suggests an inverse variation.

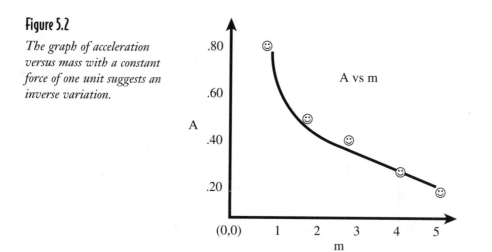

Before doing that, though, remember that there are three variables involved. Using your gray matter you may think if $A \propto \dfrac{1}{m}$ then $A \propto \dfrac{1}{m}$; that is, as m increases then A decreases by the first statement. The second statement suggests that as m increases the reciprocal of A increases and that means A must decrease. Both statements state the same relationship if this is the relationship that exists.

Use the second statement for your graph because you can introduce the third variable by substituting f for 1 since f is one unit throughout this part of the experiment. That means you can plot $\dfrac{f}{A}$ vs m on the next graph, and if a straight line emerges then $A \propto \dfrac{1}{m}$ and hence $\dfrac{f}{A} \propto m$ by your recent creative reasoning.

The graph in Figure 5.3 shows that $\dfrac{f}{A} \propto m$ and that means that $\dfrac{f}{A} = km$ and f=kmA. The constant of proportionality is k but since f and m are arbitrary units the slope of the curve is not very meaningful. Take that last expression and use the MKS system to give the constant a meaning by using units that are not arbitrary. In the expression f=kmA, using the MKS system f=(k)(kg)(m/s²) that is, mass is measured in kg and acceleration is measured in m/s². At this point a scientist might say let $k = 1 = \dfrac{bandersnatch}{\dfrac{kg \cdot m}{s^2}}$

so that when you substitute this value into the equation f=(k)(kg)(m/s²) you get f= $\dfrac{bandersnatch}{\dfrac{kg \cdot m}{s^2}}$ (kg) (m/s²)= 1 bandersnatch because when you multiply anything by one (1) you do not change its value even though you may change its looks.

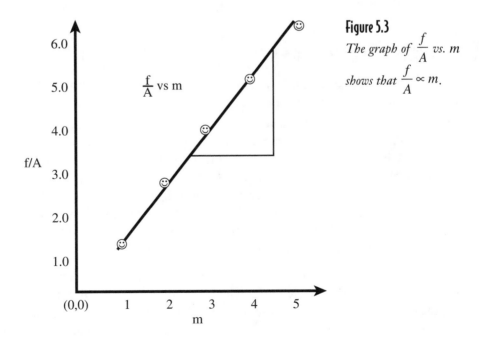

Figure 5.3

The graph of $\dfrac{f}{A}$ vs. m

shows that $\dfrac{f}{A} \propto m$.

Fortunately, physicists assigned a different value to k so that we do not have to measure force in the MKS system in bandersnatches. The assignment looks much like your suggestion because they said let $k = \dfrac{N}{\frac{kg \cdot m}{s^2}}$ so that when you substitute k into the equation you get $f = \left(\dfrac{N}{\frac{kg \cdot m}{s^2}}\right)\left(\dfrac{kg \cdot m}{s^2}\right) = 1$ newton. The assignment was made in honor of Sir Isaac Newton because of his work with force along with many other things, including his invention of the calculus.

Plain English

A **newton** is the force required to accelerate a mass of one kilogram at a rate of one meter per second squared. A **dyne** is the force required to accelerate a mass of one gram at a rate of one centimeter per second squared. A **pound** is the force required to accelerate a mass of one slug one foot per second squared.

The expression now becomes f=ma because k=1 and the force is measured in *newtons* or N in the MKS system, that is, $1N = \dfrac{1 kg \cdot m}{s^2}$. Similarly, in the CGS system $1\, dyne = \dfrac{1\, g \cdot cm}{s^2}$ and in the FPS system $1\, pound = \dfrac{1\, slug \cdot ft}{s^2}$. The pound is abbreviated lb and there is no short way of writing dyne. Out of all of that garbage you were able to retrieve a great deal of information about force. Notice that we have discussed only the magnitude of force for the most part thus far.

The Vector Nature of Force

The expression F = ma, that you developed from my experiment, is called Newton's second law. Better yet, you can identify force as a vector quantity in that equation by writing $\vec{F} = m\vec{a}$ where $\vec{F}$ is a vector, m is a scalar, and $\vec{a}$ is a vector. The equation states that if a single force is applied to an object that is free to move, the object will accelerate in the direction of the applied force with a magnitude that is directly proportional to the magnitude of the force. In all cases, the single force and the acceleration it causes act in the same direction. Soon you will be concerned with the magnitudes of those quantities. So you think a dyne is almost as bad as a bandersnatch? Worse than that is a $slug \cdot ft / s^2$, read slug-foot per square second. The hyphen (-) or dot (.), when used with units of measurement, means multiplication.

Usually you think of a push or pull as a result of objects in contact. You must keep in mind that a magnet, a planet, or an electric charge can push or pull even though two objects may not be touching. You must add forces and subtract forces just like any other vector. It may be more convenient in some cases to add forces in a somewhat different way than you have done before to find the resultant force. Some forces do push or pull by being in contact with an object. If several forces are acting on the same point in a body you may add them together two at a time by drawing a diagram that begins with their feet together. Such a diagram is called a free body diagram. Apply that idea in the solution of this problem: A sports vehicle is stuck on a muddy road. Two ropes are tied to the front bumper and a man pulls on one of the ropes with a 2.50×10^2 N force that makes an angle of 30° with a line down the middle of the road. Another man pulls on the other rope with a force of 3.00×10^2 N at an angle of 50° with the same line. What is the total force exerted on the vehicle? Both ropes are pulled parallel to the ground. Refer to Figure 5.4, which is a free body diagram of the object being pulled and the forces applied. That is, the point of application of forces and the forces themselves are all that are necessary to draw in order to solve the problem. It is not necessary to draw the men, the ropes, and the vehicle.

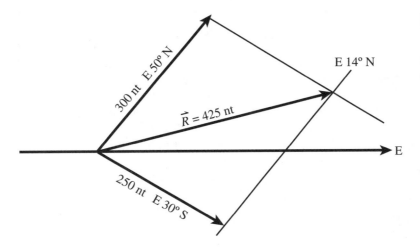

Figure 5.4

The vector diagram of forces illustrates the parallelogram method of adding two vectors.

The method of adding the forces in Figure 5.4 works only for the addition of two vectors at a time. The method of adding vectors earlier, call the *closed polygon method*, works for any number of vectors. The method used here is called the *parallelogram method*. I constructed a line parallel to one of the vectors through the head of the other. The same construction was done for the other vector. The two lines constructed in this way intersect at a point that is the vertex of the parallelogram opposite the vertex where the feet of the vectors are joined. A vector drawn from the feet of the vectors to the opposite vertex, in that order, is the resultant vector for this method. A close examination of this construction will reveal that this is just a quick way of joining the foot of the second vector to the head of the first. Remember the geometry of a parallelogram dictates that the opposite sides of a parallelogram are congruent, equal in measure. The problem was also solved by calculating the answer using a method outlined in Appendix D.

The answer is $\vec{F} = 423\,N$ at an angle of 14.4° with the line down the middle of the road. Three significant figures were used even though the original angles did not indicate that much precision. The calculated answer differs from the graphic solution by about one-half of one percent. If only two significant figures are used, the answers are exactly the same. Don't worry about the calculated answer unless you are a mathematics freak. There are some of us around with our hand-held graphing calculators and some of us even have access to a computer. Notice that, by Newton's second law, the vehicle will tend to accelerate at an angle of about 14° with the middle of the road if these are the only two forces acting on the vehicle. That may be just enough to get the SUV off the high center and free to move.

 Plain English

The **closed polygon method** of adding vectors requires that any number of vectors may be added together by joining them together foot to head until the sum is complete. The resultant vector is found by closing the polygon with a vector drawn from the foot of the first vector in the sum to the head of the last in that order. The **parallelogram method** of adding vectors requires that the feet of two vectors be located at the same point. A parallelogram is then constructed by drawing a line parallel to one of the vectors through the head of the other. A second line is constructed parallel to the second vector passing through the head of the first. The resultant vector is found by joining the feet with the opposite vertex of the parallelogram in that order.

Components of Forces

You often see people pulling or pushing an object to make it accelerate. You know that not all of the force you apply to a rope or a stick attached to the object acts to accelerate the object parallel to the ground. Only the component of the force parallel to the ground acts to do the pulling or pushing along the ground. Suppose you must apply 352 N of force to the handle of a lawn mower, and the handle makes an angle of 49° with the horizontal. How much of the force pushes parallel to the surface of the ground? Refer to Figure 5.5 for a graphic solution.

The graphic solution shows that only 231 N of the applied force acts to push the lawn mower parallel to the ground. Again the solution was calculated and the solution agrees within about two percent with the graphic solution. Notice that we resolved the applied force into two mutually perpendicular components. I promised you earlier that these two components would have important practical applications, and this is one. Were you surprised that the sum of the components is not equal to the applied force? You know that you cannot just add the magnitudes of vectors to get their sum in general.

It is not necessarily true that two plus two equals four when you are talking about vectors. Three plus three could be zero. Remember that you cannot add vectors the same way you add scalars.

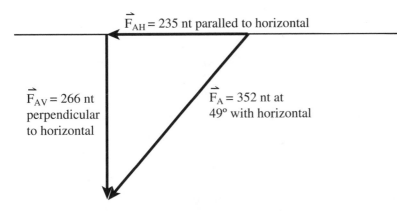

$\vec{F}_{AH}$ = 235 nt paralled to horizontal

$\vec{F}_{AV}$ = 266 nt perpendicular to horizontal

$\vec{F}_A$ = 352 nt at 49° with horizontal

Figure 5.5

The diagram of the components of a force illustrates the resolution of a force into mutually perpendicular components.

Net Force and Uniform Motion

The forces discussed so far have included component forces and total or resultant forces. You know that the single resultant force can be used to replace many forces acting on an object. Newton's second law shows what a single force tends to do to an object. Taking all of these ideas into account, the question might be what happens if we consider all forces acting on an object and arrive at one single force that can replace all of the forces acting on the object? That force is what we call the *net force*. The net force on an object can give us an indication of the type of motion an object has.

The implication on the motion of a body is that an object can travel with uniform motion only when the net force on the object is zero. Otherwise the object would follow Newton's second law and accelerate in the direction of the net force. The so-called Newton's first law enables you to explain the case for a net force of zero. Newton's first law of motion actually had its basis in the ideal experiments of Galileo in the seventeenth century. The ideal experiments were conducted only in Galileo's mind. The results of those thought experiments make up the law of inertia in Newton's first law that is paraphrased as: An object at rest will remain at rest unless a non-zero net force acts on it. An object in motion will remain in motion in a straight line at a constant speed forever unless a non-zero net force acts on it.

That means that the only way you can travel down the interstate at eighty miles per hour is to drive your car in such a way that the net force on your car is zero. It also means that when the car is brought to a sudden stop, your body tends to continue traveling at a constant speed in the direction the car was traveling initially. All of these notions imply that there are forces other than the ones you have encountered so far in this book. There are other forces that are not actually in contact with a body as well as many that are. Other forces affecting a body in motion or at rest are discussed in later chapters.

Newton's third law states: For every action there is an equal and opposite reaction. That means when you push on the wall with your hand, the wall pushes back on your hand with a force equal in magnitude but opposite in direction to the force applied by your hand. The important thing to note here is that there are two bodies involved in the application of this law. Your hand provides the action and the wall must react. Another example is your application of a force of pull on a rope attached to some object. The rope pulls back on your hand with a force that is equal in magnitude but opposite in direction to the force applied to the rope by your hand.

This law is actually an extension of Newton's second law of motion. Whenever forces act on an object, no matter what their sources may be, the motion of the object may be explained by applying all of Newton's laws. Remember that the net force is calculated by taking into account the effects of all forces on a single body. Do not worry about the force by that body on another body, Newton's third law, when drawing a free body diagram.

You may think that the men attempting to eject the mud-laden, high-centered SUV are exerting a net force, but they are not. There are other forces acting on the SUV. The only thing that we can say for the current case is that the force exerted by each of the two men can be replaced by one force that is the resultant force or the sum of the forces. Of course they have no way of replacing the two forces and that is why they pull on the silly ropes. In real life, sometimes you have to stand on the side to avoid the splatter of the mud from the middle of the road.

Suppose that you know there are only three forces acting on an object. The forces are 20.0 dynes $E15°N$, 30.0 dynes $W15°N$, and 40.0 dynes $S15°W$. What is the net force on the object? Refer to Figure 5.6 that is a graphic solution of this problem. The map convention is used to indicate direction. Two of the forces were added using the parallelogram method, then that resultant was added to the third force to get the net force. The forces were named $\vec{F}_1$, $\vec{F}_2$, and $\vec{F}_3$. The forces $\vec{F}_1$ and $\vec{F}_2$ were added to find their resultant. That resultant was added to $\vec{F}_3$ to determine the net force.

Figure 5.6 shows the final result to be 37 dynes $W54°S$ for the net force. If you calculate the net force you get 35 dynes $W55°S$. The graphic solution using the parallelogram method of adding vectors agrees with the calculated solution. That means the graphic solution is sufficiently accurate for the purposes of solving vector problems in this book. You may want to calculate solutions if you have the background or review the calculations provided in Appendix D.

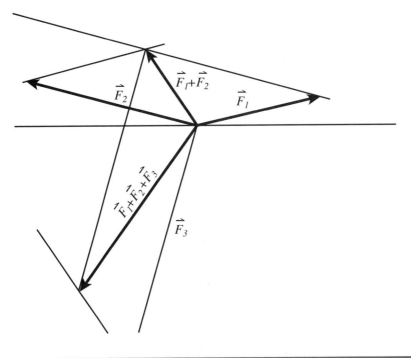

Figure 5.6

A graphic solution for the net force due to three different forces acting at a point uses the parallelogram method of adding vectors.

Physics Phun

1. How much net force must you apply to an object having a mass of 5.0 kg to cause it to accelerate at 6.0 m/s²?

2. A net force of 135 lb is applied to an object having a mass of 2.0 slugs. What is the acceleration?

3. Three forces act on a pumpkin that is placed on the frozen lake with the boy. The forces are 50.0 N North, 50.0 N *E60°S,* and 40.0 N *W30°S.* What is the net force on the pumpkin?

The Least You Need to Know

♦ Discuss how graphical analysis was used to identify the relationship between force and acceleration.

♦ Identify and use the units of force in each of the systems of measurement.

♦ Recognize the application of Newton's laws of motion to describe the motion of an object.

♦ Visualize the components of a force acting on an object.

♦ Describe the motion of an object in terms of net force.

Chapter 6

Free Fall and Gravity

In This Chapter

- ◆ Gravity, a force at a distance
- ◆ Atmosphere, air friction, and terminal velocity
- ◆ How a force at a distance affects acceleration
- ◆ What really happens when an object falls
- ◆ Weight, gravity, and force

Watching raindrops fall and splatter on the ground was a refreshing activity for me as a boy. The more rain that fell the longer we rested with no work that we could do in the fields. You have probably done the same thing, and maybe you have even filled balloons with water and dropped them to see them splatter.

Objects fall the way they do because of the force of gravity. At one time or another each of us have experienced the effects of that force in a fall. In this chapter, you will even learn to calculate the force of gravity on your body. Each of us is concerned about that force of gravity on our bodies, some of us more than others. You are about to find out how much of a pull the force of gravity has on you.

This Force Is with You

There is one force that no one can escape. It is with you from the time you pry your eyelids open in the morning until you struggle to get your weary bones into bed at night. This force has a lifelong effect on your body from before you are born until the day you are laid to rest. The *force of gravity* affects your daily activity for a lifetime.

Plain English

The **force of gravity** is the inward pull on any mass placed on, or even above, the earth's surface. This is a pull at a distance because nothing is attached to the mass pulling it toward the center of the earth.

Up to this point, there are no good explanations as to what gravity is, but we do know that it exists. Some scientists have suggested the "particle" as a good model to explain gravity and there is research in progress today at Hanford, Washington, testing the wave model as a good explanation for it (refer to Chapters 20 and 21 respectively).

Even though you may not have an explanation as to what gravity is, you know it is here and you can feel its effects in many ways. For instance, all you have to do is sit down. You feel the chair push up on your body (though some of us may experience the squashing force in this situation more than others) with a force exactly equal to the force of your body pushing downward—provided the chair is strong enough to hold you. If not, plop! You just confirm that the force of gravity exists! Your body flies straight down until it hits the floor when the chair breaks. Then, of course, the floor pushes up on you exactly equal to the force that gravity pulls down on you. If these forces are not balanced against each other, you will accelerate due to a net force. Remember from the last chapter that a non-zero net force causes an acceleration.

Notice that there is nothing attached to your body like the forces discussed in earlier chapters. Gravity is a force similar to several other forces that act at a distance. Anyone who has jumped off a diving board or fallen off a wagon or bicycle can tell you exactly what happens. Something caused them to fall downward.

Only things like birds, balloons, and boats can travel upward, remain level, or stay afloat. Of course, we use vehicles like airplanes and cars to balance the effect of gravity until we reach our travel destination. Left alone above the surface of the earth, our bodies simply fall. Only a strong parachute can reduce the effects of the fall enough to allow you to land safely on the surface of the earth. Since we do not have antigravity machines, you probably theorize that our earthly bodies are bound to the earth by gravity forever. Only a clever magician tries to convince you otherwise.

Just think how humankind has adapted to life under the influence of gravity:

◆ Men wear belts on their pants, and noses act as supports for their eyeglasses.

◆ In the home, pictures are hung from supporting wires and strong legs are constructed for tables and chairs.

◆ Tall buildings are held up by rebar, and bridges are held up by cables and supported by concrete columns.

◆ Dams are built to allow water to fall to generate our electricity.

◆ The naturally steep slopes of forested hills are efficient slides for felled trees en route to lumber mills.

◆ Elevators enable you to avoid climbing multiple stairs to gain access to the eighty-first floor.

◆ Huge, powerful rockets are required to lift astronauts and payloads into space.

These are just a few of the inventions humans have made to adapt to the world of gravity; some are advantageous and some are a nuisance.

Gravity is a force that pulls down, actually radially inward toward the center of the earth, on every object on the planet without actually touching anything. The force of gravity is measured with a *spring balance*. You cannot see gravity and we do not know what it is, but man has learned to live with its effects and to describe it as a force.

Plain English

A **spring balance** is a device containing a spring that stretches when pulled by a force. The amount of stretch is calibrated to correspond to force units, N, dynes, or pounds (units of force are discussed in Chapter 5).

Falling Objects Experience a Net Force

I was fortunate to have watched live television shots of astronauts cavorting around on the surface of the moon. They took giant leaps and landed safely in the moon dust. Their footprints recorded the telltale presence of gravity. Many experiments were conducted while they were there but none were more spectacular than those involving gravity. For example, dropping a hammer and a feather from the same height was particularly exciting. Both hit the surface of the moon at the same time! We don't observe that type of behavior near the surface of the earth. Have you ever wondered why they did that on the moon but not on the earth? That is what physics types do all the time.

They see an object behaving in a certain way and wonder if it would act that way if certain changes were made. That's the kinds of questions that this book will help you answer. What would happen if …?

Atmosphere and Air Friction

Galileo's legendary experiment of dropping two dense objects of different masses from the Leaning Tower of Pisa suggested that the objects did hit the ground at nearly the same time. Notice the difference in this experiment and the one conducted on the moon. Both objects used by Galileo were dense, not a feather and a hammer. You know the feather slowly floats to the surface of the earth while the same feather falls at the same rate as a hammer on the moon. The reason for the difference is that there is

Plain English

The **atmosphere** is the gaseous envelope surrounding the earth and held in place by gravity. **Air friction** is the force opposing the motion of an object in the atmosphere.

another vital substance present near the surface of the earth that is not on the moon: the *atmosphere.*

Suppose an object is falling near the surface of the earth. You know that gravity is pulling it downward. Another force that has not been discussed before opposes the motion of the object. Any object that is allowed to fall experiences the same force. It is the force of *air friction.* Figure 6.1 helps to visualize the two forces acting on an object, with mass *m*, falling near the surface of the earth.

Figure 6.1

The motion of an object can be described using two forces on a falling object.

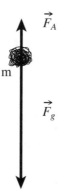

You know by viewing the vectors that the net force acting on the object is the sum of all the forces acting on the object. That means that the object in Figure 6.1 appears to have a net force acting downward. When you calculate the net force you find $\vec{F}_{net} = \vec{F}_A + \vec{F}_g$. You recognize this as a vector sum that is read as the net force is equal to the sum of the force of air friction and the force of gravity.

Analyze this statement using the conditions set by Galileo at Pisa where the object of mass m is very dense and has a relatively small surface area. That means that $\vec{F}_A = 0$, or at least so small an area it is insignificant. You can make $\vec{F}_A$ as close to zero as you like by choosing a high-density object with a small surface area.

Plain English

A **freely falling body** is an object that is influenced by the force of gravity alone.

You can see that the net force on such an object is $\vec{F}_{net} = \vec{F}_g$. Note that this is a very special case in the analysis. The object is said to be a *freely falling body* in this case.

Terminal Velocity

Suppose the falling object is not dense, say a ball of newspaper. The air friction is not zero and the net force is less than the force of gravity. Even if the object is dense but has a relatively large surface area, the net force is less than the force of gravity. For instance, when a skydiver jumps from a plane and falls, her body can reach speeds of nearly 120 mi/hr when she falls from a high altitude. At first her speed downward is zero, but as soon as she is out of the plane she starts to accelerate downward because of the net downward force on her. Her downward speed increases until she reaches her top downward speed. That maximum speed is called the *terminal velocity*.

Plain English

Terminal velocity is the maximum velocity of a falling object in air. The air friction on a falling body will increase with velocity until the upward drag of air friction is equal in magnitude to the downward force of gravity on the body. At that point there is no net downward force, and its downward speed no longer changes.

A terminal velocity can be achieved if she holds her arms and legs close to her torso, decreasing the surface area exposed to the atmosphere. Stretching out her limbs to increase the surface area can reduce her speed and enable her to change horizontal direction. Fortunately, the parachute she jumps with can reduce the speed to about 14 mi/hr before she touches the earth. It is helpful to be able to calculate the speed of a dense object falling freely above the surface of the earth because that will give us insight into falling objects like the human body.

If you wonder why a relatively dense object reaches a terminal velocity, think about what is taking place in terms of Newton's laws. You may reason that a terminal velocity implies zero acceleration. You are correct. How can a body have zero acceleration? You recall that Newton's first law applies to that situation and that means that there is no net force acting on the body. That means that the force of gravity must be equal in magnitude but opposite in direction to the force of friction. Your explanation would conclude that a falling body that is relatively dense compared to the amount of surface area exposed to the air will experience an increase in air friction with increased speed. When the force of friction reaches the magnitude of the force of gravity, the sum of the two will be zero and the object will travel with a constant velocity called the terminal velocity.

Acceleration Due to a Force Acting at a Distance

You see that an object falling freely experiences a considerable increase in speed. You know that means the object accelerates. The acceleration must be due to the net force according to Newton's laws of motion. Since the net force is the force of gravity for dense object with small surface areas, the acceleration of the object must be due to the force of gravity. That is, since $\vec{F}_{net} = \vec{F}_g$, then $m\vec{a}_{net} = m\vec{a}_g$. Because the object is dense and does not gain or lose mass as it falls, $\vec{a}_{net} = \vec{a}_g$.

You probably suspected that the net acceleration is the same as the acceleration due to gravity under the given circumstances—that is, one in which the drag due to air friction is nonexistent. The acceleration due to gravity is so important that it is assigned a special symbol of its own. The acceleration due to gravity is represented as g with a direction radially inward toward the center of the earth or a direction downward as you would describe it locally. Not only does the acceleration due to gravity have a special symbol assigned to it, but there is also a special value for it.

Experiments measuring the rate at which objects speed up when there is "no" air friction have shown that near the surface of the earth $g = 9.80 \frac{m}{s^2}$ in the MKS system, $980 \frac{cm}{s^2}$ in the CGS system, and $32 \frac{ft}{s^2}$ in the FPS system (refer to the section "Making Measurements" in Chapter 1). This means that if an object fell for a second near the earth, it would be traveling at a speed of 9.8m/s downward. If it fell for two seconds, it would be traveling at twice that speed. The convention that is used to indicate the direction of the acceleration of gravity is to say that the acceleration due to gravity is $-g$, which means the opposing force of air friction on a falling body is positive. That is, for motion near the earth's surface the convention used is: up is positive and down is negative.

Newton's Figs

Remember that the terms velocity and speed are interchangeable only when motion is along a straight line in one direction.

Describing the Motion of a Falling Object

You are able to describe the motion of a falling object by using the information just developed. You know that if the opposing force of friction influences the object, the object can fall long enough to reach a terminal velocity and will then travel with uniform motion.

If you assume an ideal condition where the object is not influenced by air friction, then the motion is uniformly accelerated motion. The acceleration is the acceleration due to gravity that is constant near the earth's surface. That means that you can use all of the information developed in earlier chapters to describe the motion of a falling object as either uniform or uniformly accelerated.

 **Newton's Figs**

Whenever you read a physics problem and find that an object is dropped or is initially at rest then you are given that $v_0 = 0$ m/s.

As soon as you read a problem that involves an object falling near the surface of the earth, you will know that the acceleration is $-g$. Also make note of the initial conditions of the motion. If the object is dropped, that means that the initial velocity is zero.

The description of an object falling freely will involve three major ideas:

♦ displacement

♦ average velocity

♦ acceleration

Let the magnitude of displacement be represented by y since the motion is either up or down. This is a convention borrowed from graphing exercises in algebra where the ordinate of a point, y, is plotted by moving vertically and the abscissa, x, is plotted by moving horizontally. Using ideas developed in earlier chapters:

♦ distance is calculated as $y = \bar{v}t$

♦ average velocity $\bar{v} = \dfrac{v_f + v_0}{2}$

♦ acceleration is $-g = \dfrac{v_f - v_0}{t}$

If you feel that you must memorize information, memorize these three ideas for the description of uniformly accelerated motion. Better yet, you might just think about these quantities and recall that distance is simply average rate times time, average speed is the sum of initial and final speed divided by two, and acceleration is the difference in velocities divided by the time required for the velocity to change.

Problem:

Suppose you are asked to find the distance traveled by a falling object if you are given initial velocity, time, and acceleration. The algebraic solution is outlined here so that you can use that solution as a model for solving algebraically all problems of this type. Find y given v_0, $a = -g$, and t.

Solution:

$y = \bar{v}t$ Begin your solution with the quantity you are solving for expressed in terms of at least one quantity that you are given or that is implied by the statement of the problem.

$y = \left(\dfrac{v_f + v_0}{2}\right)t$ Substitute for a quantity you are not given in terms of at least one quantity that is given or implied.

$-g = \dfrac{v_f - v_0}{t}$, so $v_f = v_0 - gt$ Substitute this into second step above.

$y = \left(\dfrac{v_0 - gt + v_0}{2}\right)t$ After substitution.

$y = \left(v_0 - \dfrac{gt}{2}\right)t$ Simplifying.

$y = v_0t - \dfrac{gt^2}{2}$ or $y = v_0t - \dfrac{1}{2}gt^2$. This expression is used later to discuss the motion of objects thrown into the air from the surface of the earth. In that case, y would be the distance above the surface of the earth and not the distance the object falls as in the next example problem.

Problem:

Suppose an object is dropped from a point above the surface of the earth and falls for 3.0 s. How far does it travel?

Solution:

The algebraic solution was constructed earlier, $y = v_0t - \dfrac{gt^2}{2}$. The specific solution for this problem is $y = (0)(3.0s) - 32\dfrac{ft}{s^2} \times \dfrac{9s^2}{2} = -144\,ft$. That is, you know that the object fell 144 feet.

Problem:

Solve algebraically for y if you are given v_f, v_0, and a=−g. Using the same outline as before, you may begin with the calculation of y in terms of at least one thing that is given directly or indirectly.

Solution:

$$y = \bar{v}t$$

$$y = \left(\frac{v_f + v_0}{2}\right)t \quad \text{Using the definition of average speed.}$$

$$y = \left(\frac{v_f + v_0}{2}\right)\left(\frac{v_f - v_0}{-g}\right) \quad \text{Using the definition of acceleration and substituting.}$$

$$y = \frac{v_f^2 - v_0^2}{-2g} = \frac{v_0^2 - v_f^2}{2g} \quad \text{Deduced by multiplying and then simplifying the algebraic expressions.}$$

Suppose an object is dropped from a point above the surface of the earth and reaches a velocity of $12\frac{m}{\sec}$. How far has it traveled?

You have just completed the algebraic solution to this problem. The solution to this specific problem is $y = \dfrac{-\left(12\frac{m}{s}\right)^2}{2\left(9.80\frac{m}{s^2}\right)} = -7.35m$. It has fallen 7.35 m.

It is easier to keep track of the details of a solution to problems like these if you solve the problem algebraically first. Then substitute the numerical values into the algebraic solution to find the numerical solution.

Did you notice that $v_0 = 0$ in each of these problems? You know that $v_0 = 0$ because the object was dropped in each exercise. Also notice that we did not calculate the height of the object in either case. We found only how far it traveled vertically downward after having been dropped. That is why the numerical answer is negative, the object is displaced downward.

Physics Phun

Solve each problem algebraically first and then substitute numerical values into the algebraic solution and calculate the numerical answer.

1. How many feet does a dense object travel in the first second if it is dropped from a point above the surface of the earth? How far does it travel in the second second? How far has it traveled in two seconds?

2. If a stone is dropped from a point above the surface of the earth and reaches a speed of 20.0 m/s, how far did it travel? How long was it falling?

Mass vs. Weight

You calculated the net force on a falling object near the earth's surface. Then you calculated the acceleration of a falling object when the force of air friction is negligible. When we made that latter calculation we divided both sides of the equation by m as reviewed here:

$$\vec{F}_{net} = \vec{F}_g$$
$$m\,\vec{a}_{net} = m\,\vec{a}_g$$
$$\vec{a}_{net} = \vec{a}_g$$

This argument and result are correct, but you should be aware of the fact that it appears there are two types of mass:

♦ gravitational mass, which is determined by using a device such as a *triple beam balance*

♦ inertial mass, which is determined by a different type of balance called an inertial balance, which measures the amount of opposition to motion an object has due to its inertial mass

Plain English

A **triple beam balance** is a device for measuring mass by comparing an unknown mass with a known mass by balancing two pans. The two pans are attached to a beam that is supported by a fulcrum in much the same way as a seesaw or teeter-totter.

You know that the more mass an object has, the harder it is to stop it when it is in motion. Similarly, the more inertial mass an object has, the more difficult it is to start it in motion. It is important to note that inertial mass and gravitational mass are directly proportional. Furthermore, if you choose to measure both with the same unit of mass, they are equal. That means they are measures of the same quantity. You may wonder why it is important to make the distinction between the two. There would be no need if we could always measure mass with the same instrument. That is not the case with mass.

Suppose you want to measure the mass of an object in a spaceship orbiting the earth. A triple beam balance does not work in the spaceship. We can use an inertial balance calibrated to measure inertial mass in kilograms to determine the inertial mass of the object. One inertial balance is essentially a physical pendulum calibrated to measure inertial mass by the length of its period. The longer the period, the greater the inertial mass. After we develop these ideas about mass in this chapter, we will remember that inertial mass and gravitational mass are directly proportional. Measure both types of mass in the same fundamental unit, then refer to just the mass of an object.

Weight, the Force Due to Gravity

The force due to gravity has an everyday name that you might be all too familiar with. Many of you probably exercise daily to deal with that force, known as *weight*. Make note of the fact that weight and mass are two entirely different things. Weight is a force due to a gravitational field that you are in, like the earth's field, and is therefore a vector quantity. Mass is a scalar (see Chapter 3 for more information on vectors and scalars). Weight is a derived quantity and mass is a fundamental quantity. Both are measures of matter, but they are entirely different measures.

Newton's second law shows us that mass and weight are proportional but not equivalent. Using that law we can state the relationship symbolically as $W = m_g g$. The statement reads as: weight is equivalent to the product of gravitational mass and the acceleration due to gravity.

Plain English

Weight is the force of gravity on any object. It is a vector quantity with direction radially inward toward the center of the earth. The direction as described locally is down.

You can see that mass and weight are proportional but not equal, and the constant of proportionality has the value of the acceleration due to gravity. The value is also called the earth's gravitational field constant. It has units of N/kg. One kilogram has a weight of 9.8 N here on the earth. That is the amount of force with which the earth pulls on a kilogram. Of course, that is the force with which the kilogram pulls on the earth also. Can you think why the earth does not accelerate toward the kilogram as much as the kilogram accelerates towards the earth? That's right, the earth is much more massive. You can see that if you were to multiply mass in kg times 9.8 N/kg, you would get an answer in newtons. If you are given the gravitational mass of an object, you are able to calculate the weight of the object. Also, if you are given the weight of an object, you can calculate the gravitational mass of an object.

Weight is a derived quantity and you know to check the units of any new quantity when it is introduced. Again, using Newton's second law, you find that $W = m_g g$ means that in the MKS system W is expressed in $\frac{kg - m}{s^2}$ or N. In the CGS system W is expressed in the units $\frac{g - cm}{s^2}$ or dynes; also, in the FPS system W is expressed in the units $\frac{slug - ft}{s^2}$ or lb.

Often you can check your solution to a physics problem by checking the units of your answer. For instance, if you are calculating the weight of some object and you find that the units are $\frac{slug - ft}{s}$, then you know that you have either made a mistake in algebra, unit analysis, or substitution. It may be that you are using the wrong idea or maybe even solving the wrong problem. The wrong units suggest an error, so go back and check all of your work.

Physics Phun

1. Your physics teacher has a mass of 5.0 slugs. How much does he weigh?
2. The athletic trainer weighs 490 N. What is her mass?

Gravitational Field

You notice that I have emphasized that the force of gravity discussed so far has been a force near the surface of the earth. The reason is that the force on a one-kilogram mass near the surface of the earth is about 9.80 N. It is correct to say about 9.80 N because the weight of a one-kilogram mass at the North Pole is nearly 9.83 N, and it weighs about 9.78 N at the equator. You see that the force of gravity on a unit mass varies slightly depending on where on the surface of the earth it is measured. Even though the magnitude of the force of gravity on a unit mass varies a bit depending on the location on the surface of the earth, the direction remains radially inward toward the center of the earth.

It does not matter where you measure the force of gravity, on the earth's surface or above it; the direction is always radially inward toward the center of the earth. Above the surface of the earth, the force on a unit mass varies from point to point as you move away from the earth's surface.

No matter where you place a unit mass above the surface of the earth, the force of gravity pulls on the mass in a direction back toward the earth. To help you to visualize what is going on here, imagine that the surface of the earth is just one of many spherical surfaces having the center of the earth as their common center. If you measure the magnitude of the force of gravity on a unit mass at each point on any one surface, you find that the measurements are the same. Even though that is correct, you find a different reading for each surface. In fact, the measurement of the force of gravity on a unit mass becomes smaller as you move away from the center of the earth. The situation just outlined is what physicists call a *gravitational field*.

Plain English

The earth's **gravitational field** is that region of space where the force of gravity acts on a unit mass at all locations on or above the surface of the earth.

Later you will calculate the magnitude of the earth's gravitational field at any altitude above the surface of the earth. It is important at this time to be able to calculate the magnitude of the gravitational field at any point given the force of gravity on a unit mass. You know that Newton's second law enables us to calculate the force of gravity.

$$\vec{F}_g = m_g \, \vec{g} \quad \text{By Newton's second law.}$$

$$\vec{g} = \frac{\vec{F}_g}{m_g} \quad \text{Dividing both sides of the previous equation by m.}$$

If you use a unit mass in the last equation, you will calculate the force of gravity on a unit mass that is the magnitude of the gravitational field. The vector $\vec{g}$ represents the gravitational field at any point. Remember that the direction of that vector is radially inward toward the center of the circle as indicated in Figure 6.2 for a couple of altitudes above the surface of the earth.

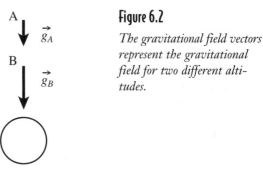

Figure 6.2

The gravitational field vectors represent the gravitational field for two different altitudes.

Imagine concentric spherical surfaces with the vectors $\vec{g}_A$ and $\vec{g}_B$ at every point on the surface and you will get an idea about the gravitational field of the earth for two different altitudes. The units of the gravitational field in each system are: $\frac{N}{kg}, \frac{dynes}{g}$, or $\frac{lb}{slug}$.

Even though I used only two altitudes, be aware that any mass placed anywhere in the region of space surrounding the earth will be affected by the gravitational field. A force at a distance was used early in this chapter to describe the force of gravity to prepare you to accept the notion of a gravitational field.

Force Field

The idea of a force field is important to understanding several different ideas in physics. For example, there are electric fields and magnetic fields. Probably you are familiar with both. The source of the force as well as the unit particle that mediates them is different from gravity but the basic concept of field is the same. If you have ever been caught out in a thunderstorm and you felt the hairs on your body stand up, you knew that you were in an intense electric field. You were probably about to be struck by lightning so you took cover!

Usually there are two or more bodies involved in the definition of a field, two charges or two magnetic poles, for example. The earth was one body defining your thunderstorm and a charged cloud above the earth was the other. You were lucky enough to be a particle affected by the electric field. Later you will find that the earth is not the only body involved in the definition of the gravitational field of the earth. You can think of all the people on earth as being little particles in the gravitational field of the earth.

The Least You Need to Know

◆ Recognize how mass can be measured using a platform balance in a gravitational field or by an inertial balance.

◆ Distinguish mass from weight.

◆ Identify the units of weight in each system of measurement.

◆ Calculate the weight of an object given its mass.

◆ Explain what a freely falling body means.

Chapter 7

Projectile Motion

In This Chapter

- ◆ Horizontal motion of objects thrown near the earth
- ◆ Vertical motion of objects near the earth
- ◆ Both vertical and horizontal motion simultaneously
- ◆ The flight of a dense object
- ◆ Baseballs and footballs

You have already considered falling bodies and how they are affected by air friction. We continue with that discussion and include the force of gravity, an old friend by now. By the end of this chapter we will combine all of that information along with our ability to describe motion to explain the motion of an egg that is sent flying by an egg launcher, a potato from a potato gun, or a bullet from a hunter's rifle. All of these objects behave in about the same way when they travel through the air near the earth's surface.

For our explanations we will consider small sections of the earth to be flat. We can take into account rolling hills or the fact that the earth is really curved when we have mastered the simple case. All of the items mentioned previously are classified as projectiles, just like the spit wads that flew past my ear and stuck to the chalkboard while I was drawing diagrams for my students. They did appreciate my efforts but agreed with me that there is nothing like some good old laboratory experiments to bring physics to life. You might try some of the experiments I suggest from time to time throughout this book. Not with spit wads, though.

Effects of Forces Near the Earth

A force near the surface of the earth that affects the motion of every object is a familiar one, the force of gravity. We may ignore as insignificant the force due to air friction by choosing to work with a dense object. The first motion I'll describe is what you would observe if you dropped a dense object. I used to ask for volunteers for a dense object to drop. The class invariably volunteered me as the densest object in the room. You think I'm joking? I once had a young man volunteer to be a simple pendulum. Two husky classmates stood on my classroom demonstration table and held his legs while he swung back and forth, establishing the self-proclaimed simplest pendulum ever.

Suppose you drop a stone from a bridge and three seconds later you see it splash. What is the height of the bridge? We established in an earlier chapter that dropping an object means the initial velocity is zero. In an analysis along with that information, we found that $y = -\frac{1}{2}gt^2$. The distance the stone fell is then $y = -\frac{1}{2}\left(32\frac{ft}{s^2}\right)\left(3\,s^2\right) = -100$ ft. We are allowed to keep only one significant figure in our answer according to the statement of the problem. The bridge must then be about 100 feet above the water to a precision of one significant figure. Any time you know the acceleration due to gravity and the time an object is in the air, you can calculate the distance it moves vertically above the earth using $y = v_0 t - \frac{1}{2}gt^2$ as derived in an earlier chapter. An important concept to consider is that if the object is in motion horizontally above the earth, it travels vertically (due to the effect of gravity) as well as horizontally.

A Johnnie's Alert

Remember to check the units of measurement of all quantities involved in a problem and the number of significant figures in each measurement.

Moving On, Across Only

You solve crossword puzzles with moves down and across. Concentrate on motion horizontally or across first. I choose to discuss this motion to begin with because it is the simplest type of motion to describe; that is, if the object is dense so that we can focus on ideal motion horizontally. Under these conditions air friction does not play a significant part in our description of motion. Of course air friction affects any motion in the atmosphere of the earth. That is why I say the motion is ideal, so that you will be aware that there is no net force on the object acting in the horizontal direction while it is traveling horizontally in the air. There are other motions involved, such as the rotation of the air with the earth and the movement of objects toward the earth, but for now we will deal only with the horizontal motion of an object.

Suppose we launch an object horizontally with an initial horizontal velocity. In order to distinguish this velocity from any other you may encounter in this discussion, I label its magnitude v_{0H}. Since this is the motion of an ideal object, the force of air friction is negligible or zero. Therefore, the motion of this object horizontally is uniform with constant velocity magnitude, v_{0H}.

I borrow the algebraic graphing techniques again and call the horizontal displacement x, as I have done in earlier chapters. The magnitude of the horizontal displacement is calculated as $x = v_{0H}t$ where t is the time the object is in motion horizontally. In Figure 7.1, the path of the object is plotted with the two vectors representing its horizontal velocity as the object moves along the path. The horizontal velocity vector tangent to the path shows the velocity at the instant of launch. Recall that the motion of an object in a circular path at a constant speed has the velocity vector tangent to the path at every point in the path. The velocity in both cases is called the *instantaneous velocity*.

Plain English

The **instantaneous velocity** is the velocity of an object at any instant of its motion. The instantaneous velocity vector is tangent to the path of motion of the object at every point along the path.

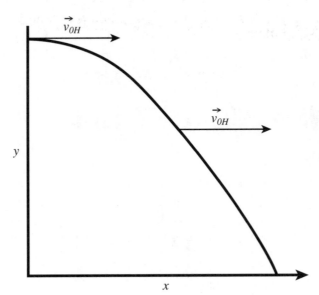

Figure 7.1

The path of an object near the surface of the earth launched horizontally.

The horizontal velocity in this case is the instantaneous velocity only at the time of launch. At all other points along the path of the object in Figure 7.1, it is the horizontal component of the instantaneous velocity. Usually when you refer to the velocity of an object in motion, it is understood that you are referring to the instantaneous velocity unless you specify a component of the velocity or the average velocity.

It may seem that I am being very picky. However, I am giving you some information that will help you to understand the meaning of your textbook or some other reference on physics. Do you remember the problem I posed for you about the runner holding a steel ball? The runner dropped the ball while maintaining his speed in a straight line. Well, now you know the answer if you did not figure it out for yourself. The ball will land at his feet the same distance away from his body, as it was when he first dropped it. The path of the steel ball would be the same as the one in Figure 7.1. "Why?" you ask. Because the steel ball has negligible force due to friction acting on it so it travels with uniform motion horizontally. In fact, the horizontal velocity of the steel ball stays the same as the runner, assuming that he does not change his velocity. That is why both are at the same location when the ball hits the ground.

Consider for a moment a ball that is not dense launched horizontally into the air. For instance, when you attempt to describe its motion, a ping-pong ball is far from ideal. You know what the path should be if air friction did not affect it. It would be just the same as the steel ball dropped by the runner. If the ping-pong ball is launched just like the ideal ball traveling from left to right in Figure 7.1, you could make the path much longer or much shorter than that path. Launching the ping-pong ball with a counterclockwise spin can make it longer, while a clockwise spin can make it shorter. In either case, the horizontal velocity does not remain constant so our description of the motion of the ball in Figure 7.1 does not hold for a ping-pong ball. For now, we concentrate on the motion of a relatively dense object and use its ideal path to guide us in our thinking about less dense objects when the need arises.

What Happens When an Object Is Thrown Vertically Upward

We have discussed dense objects dropping and traveling horizontally in the atmosphere of the earth. You know that an ideal object launched horizontally will travel at a constant speed if we ignore the effects of gravity. The force of gravity will accelerate an object dropped from a point above the surface, and ideally it will travel with uniformly accelerated motion.

Keeping all of that information in mind, consider the motion of an object under ideal conditions when it is launched straight up. Before we can describe the motion we must identify physical conditions at launch time, upon returning to the earth, and for the total amount of time it is actually in the air. State precisely what the problem is and what information the statement of the problem gives.

You can identify an important piece of information or two by thinking about the initial conditions of the problem. In this case, it is not what is given explicitly, but what the statement implies. If the object is launched, what is implied? Of course, it must be given an initial velocity upon launch. Not only that, but the initial velocity is vertical only and directed upward. As soon as it is launched, your mind tells you something about a situation like this.

Newton's Figs

Many times I find that I actually must read the textbook and the problem at hand with a pencil and paper. That is, I focus on what the problem states, then write down what is given or implied and what I am to find. For those of us who may serve as an ideal object to be dropped in the experiment, that is a good plan.

Think now: If it goes up and comes back down, what must happen along the trip? Yes! Somewhere up there it stops, turns around, and returns to the earth. You don't think it stops? Good, I like that skepticism. Don't let me or anyone else lead you down a blind path. Keep everyone honest if they are to communicate with you. It doesn't stop for even a fraction of a second? You are correct. It stops for only an instant. How long of a time is an instant? Now that is a good question. An instant is shorter than the smallest time that you can think of. I know you think that the object doesn't stop. However, it must stop if it comes back down. We know that it has a net force on it; otherwise it would continue to travel in a straight line forever.

If it stops up there somewhere, what can you say about the distance from the surface of the earth? What is the acceleration at that point? Do these questions occur to you when you are thinking about solving a problem? I hope so, but if not, make note of all of these ideas and see if they help us to formulate a plan for a solution. This process of critical thinking is the important thing we should keep in mind when studying any discipline, but especially physics. Keep in mind that it is not absolutely true that the object will return to the earth because, for example, a giant bird may grab it and fly off to some distant nest.

Based on our experience, and the physics we know at this point, we can feel confident that the questions we have raised are good ones. You can almost taste a solution brewing, as well. What is the acceleration when the object stops at the peak of its trip upward?

A Johnnie's Alert

Discussing motion along a vertical straight line involves vector quantities. However, the convention that is used to indicate the direction is a plus sign for motion upward and a minus sign for motion downward. That means the vector will be properly represented with the magnitude of the vector and the appropriate algebraic sign.

No answer yet? Let that wait and we will begin to write down in symbols the answers to questions we have in mind as well as the information either given or implied.

Let the initial velocity of launch be represented by v_{0V}, which means the initial vertical velocity or the vertical velocity when time is zero or when we start measuring time. The net force is the force of gravity, so the acceleration is $-g$. When the object rises and then stops, that gives us a very important piece of information that we can label $v_{fV} = 0$ and understand it is for the trip upward. Represent the height above the surface of the earth as y. That means $y = 0$ initially or symbolically when $t = 0$. We have enough information to describe the motion:

$y = \bar{v}t$ A good place to begin for uniformly accelerated motion.

$y = \left(\dfrac{v_{0V} + v_f}{2} \right)t$ By substitution.

$y = \left(\dfrac{v_{0V} + v_{0V} - gt}{2} \right)t$ From the definition of acceleration and substitution.

$y = v_{0V}t - \dfrac{1}{2}gt^2$ Simplifying, then multiplying by t.

This final expression will give us the height of the object at any time t because we will know the initial velocity and the acceleration of gravity. Big deal, you say? Maybe some solutions to specific problems will help to show what we have accomplished with this solution.

Physics Phun

1. A stone is thrown straight up with an initial velocity of $5.0 \times 10 \, \dfrac{mi}{hr}$. How high is it in 1.0 s? In 2.0 s? In 3.0 s? What is its maximum height above the surface of the earth?

2. A steel ball is thrown upward with an initial velocity of $1.00 \times 10^2 \, \dfrac{km}{hr}$. What is its maximum height? How high is it in 3.0 s?

Two Motions: Uniformly Accelerated (Vertical) and Uniform Motion (Horizontal)

You are prepared to tackle one of the most difficult problems in a first course in physics. I will state a problem that you recognize immediately as complicated. Then you will analyze the problems into parts, solve the parts, and then synthesize the solutions to the parts to construct a solution to the complex problem. This is a problem-solving strategy that will work on the solution of many problems including but not limited to physics, mathematics, and chemistry.

The solution to our physics problem is not easy using algebra because of some intuitive leaps that will be used. If you use calculus the problem is a snap, a good reason to include calculus in your educational plan. I will stick with algebra here and guide you through the sticky parts. I won't call those parts difficult, because it depends on your mastery of ideas discussed in this chapter. You will be successful and pleased with yourself for making the effort.

After completing this section, you will be able to throw an object into the air and know where it lands. You may think that is no big deal because we have just done that. However, what if you launch an object into the atmosphere at an angle, not just straight up? The fun part of the solution is that it appears there is not enough information given to solve the problem. Sometimes that seems to be true of a lot of problems in physics, but if you think through the problem (as we did with the vertical launch) you will surprise yourself at the creative solutions you are able to construct. Correct solutions, of course!

Understanding Projectiles

From this time forward, I refer to the ideal object launched above the surface of the earth as a projectile. I know that is a depressing term, but just think potato, water balloon, egg, or ripe tomato. The problem you want to solve involves a projectile. A projectile is launched into the atmosphere with an initial velocity v_0. You are to find the *time of flight*, *maximum height*, and *range* of the projectile. Let T represent the time of flight, Y the maximum height, and X the range. All of these quantities are labeled in Figure 7.2. All of the quantities needed to solve the problem are labeled in that same figure.

Plain English _____

The **time of flight** of a projectile is the total time it is in the air. The **maximum height** of a projectile is the greatest distance the projectile is above the earth or the level of launch. The **range** of a projectile is the maximum distance traveled horizontally by the projectile. It is measured from the point of launch to the point of return to the same level.

Figure 7.2

The path of a projectile launched at or near the earth's surface.

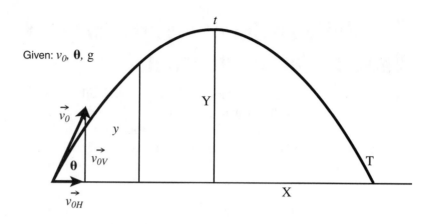

Given: v_0, θ, g

Find: T, Y, X

Also find the information given and the information you are to find listed in the upper left and lower left of Figure 7.2. Is it clear to you that we are given all of the information we need to solve the problem algebraically? I will guide you through the solution step by step.

The most difficult part of this problem is realizing that when you are given $\vec{V_0}$, you are given a treasure of information as a student of physics. That vector not only gives you the magnitude v_0 but also the angle, θ, at which the projectile was launched. You also know that the acceleration is –g because the object is in motion above the surface of the earth. The analysis begins with the realization that the projectile is undergoing two types of motion simultaneously: uniform motion horizontally and uniformly accelerated motion vertically. That gives you a plan to follow as a solution. Describe the horizontal motion of the projectile first and then describe the vertical motion. After you are satisfied with those solutions, synthesize those parts to construct a solution to the complex problem as first stated.

Describing the horizontal motion requires that we are given the initial horizontal velocity. We can deduce the initial horizontal velocity from the initial velocity by using an idea developed in the section "Components of Vectors" in Chapter 3. We must resolve the initial velocity into two mutually perpendicular components, one vertical and one horizontal. You see in Figure 7.2 that I have done that for you already.

You know that the initial velocity is the instantaneous velocity at the point of launch, so it is tangent to the path of the projectile at that point. I constructed the angle θ that is given as the direction of v_0 along with the projections of the initial velocity on the horizontal and vertical axes. That was accomplished by dropping a perpendicular from the head of the initial velocity vector to the horizontal.

The geometry of the constructed right triangle dictates the magnitude of the horizontal component of the initial velocity. I realized that since the opposite sides of a parallelogram are congruent, equal in measure, the vertical component is equal in length to the perpendicular just constructed. I labeled the initial horizontal component v_{0H} and the initial vertical component v_{0V}. Remember that these are vectors acting along straight lines so direction is indicated with a plus or minus sign. You must do all of this for yourself when you solve projectile problems on your own, and remember to do the following:

- Construct the given angle θ with a protractor.

- Construct the vectors to scale so that you can measure the magnitudes of the components as part of the information given in the statement of the problem.

You are ready to dive into this mud puddle.

Launching a Projectile Vertically Upward

You know that you must know the time the projectile is in the air to completely describe its horizontal motion. You used that idea earlier when the projectile was launched horizontally. Now think of the projectile that was thrown vertically upward. The projectile went up, stopped, and turned around and came back to earth, barring a capture by a hungry bird. You see that the object was slowed and stopped by the same thing that speeded it up on the way back down, the acceleration due to gravity. Do you see that can happen the way it does only if it takes just as much time to travel from the launching point to maximum height as it does to travel from that point to the same level from which it was launched? Let the time required to reach maximum height be represented by t.

You may wonder what that little t is doing at the top of the path in Figure 7.2. I put it there to remind me that it takes that time to get there no matter whether it goes straight up or along the curved path as shown. You are probably well aware by now that the horizontal motion and vertical motion are independent of each other. It takes the projectile just as long to travel from that point to the landing point where you will see T on the diagram. I put that there to remind me that the projectile is in the air for a time T. I realize that $T = 2t$. That means that the time of flight of the projectile, T, is twice the time required to reach maximum height. Really no one needed to tell you that because you figure that information by training your mind to think that way.

Plain English

A dense object launched near the surface of the earth at a low initial velocity will undergo two types of motion simultaneously, and they are independent of each other. Vertically the motion is uniformly accelerated because of the acceleration due to gravity, and horizontally it is uniform motion.

Notice that you are slowly but surely gathering all the clues you can from the information given so that you can solve the mystery. There is a small problem here if we are to describe the horizontal motion. We have used the symbols t and T to represent two specific times. Let τ be a symbol to represent any time the projectile is in flight. We can describe the horizontal motion now, as you can see. Before we do, go back and remind yourself what the problem is again. You are given only the initial velocity and the acceleration due to gravity. You are to find the time of flight, the maximum height, and the range of a projectile.

Did you notice that in the process of thinking through the description of the horizontal motion you discovered a solution for another part of the problem? That is the time of flight is twice the time to reach the maximum height! A discovery like that is called a discovery by serendipity. Isn't that a beautiful word? It is an exciting process as well, in which you look for a solution to one problem and by serendipity discover a solution to another. Several scientific discoveries have been made that way; the discovery of Teflon is one good example. The mud puddle is beginning to allow us to see clearly now.

We know that $T = 2t$. The horizontal motion can be described with the expression $x = v_{0H}\tau$, that is, the distance traveled horizontally is the horizontal component of the initial velocity multiplied by the time. That means that the distance traveled horizontally for half the trip is given by $x = v_{0H}t$, and for the whole trip the range is $X = v_{0H}T = v_{0H}(2t) = 2v_{0H}t$! Our problem is two thirds solved. Consider the third part of the solution, and when it is complete I will summarize all of our work. The vertical motion is uniformly accelerated motion because the net force on the projectile is the force of gravity. You can describe that motion just as we did earlier:

$y = \bar{v}\tau$ For the height at any time it is in the air.

$y = \left(\dfrac{v_f + v_{0V}}{2}\right)\tau$ Substituting the definition of average velocity.

$y = \left(\dfrac{v_{0V} - g\tau + v_{0V}}{2}\right)\tau$ Substituting using the definition of acceleration.

$y = v_{0V}\tau - \dfrac{1}{2}g\tau^2$ By multiplying and simplifying.

You have developed a solution for the height at any time the projectile is in the air. The maximum height is the height when $\tau = t$ so $Y = v_{0V}t - \frac{1}{2}gt^2$.

A summary of our solution so far includes:

1. $T = 2t$ The time of flight is twice the time to reach maximum height. You will calculate the time to reach maximum height later.

2. $X = 2v_{0H}t$ The range is twice the product of the horizontal component of initial velocity and the time required to reach maximum height.

3. $Y = v_{0V}t - \frac{1}{2}gt^2$ The maximum height is the product of the vertical component of the initial velocity and the time required to reach maximum height less one half the product of the acceleration due to gravity and the square of the time required to reach maximum height.

Aren't you glad you can use algebraic symbols and do not always have to express these ideas in words? I used words to emphasize that we have solved the problem in terms of the time required to reach maximum height. The problem will be solved only when we have expressed all three of the parts of the analysis in terms of just the information given: v_{0V}, θ, and $-g$. You need to find the time required to reach maximum height in terms of one or more of the quantities given. There is a piece of information that is implied as we found from an earlier analysis, and that is the final vertical velocity at maximum height; the projectile stops moving vertically at that point. You know v_{0V}, $-g$, and $v_{fV} = 0$. Use the definition of acceleration to solve for the time required to reach maximum height.

$-g = \dfrac{v_{FV} - v_{0V}}{t}$ The definition of acceleration.

$t = \dfrac{v_{0V}}{g}$ Multiplying and simplifying.

Now we can solve the problem by substituting the value for the time required to reach maximum height into all three of our partial solutions.

1. $T = 2t$, so $T = 2\left(\dfrac{v_{0V}}{g}\right) = \dfrac{2v_{0V}}{g}$.

2. $X = 2v_{0H}t$, and $X = 2v_{0H}\left(\dfrac{v_{0V}}{g}\right) = \dfrac{2v_{0H}v_{0V}}{g}$.

3. $Y = v_{0V}t - \dfrac{1}{2}gt^2$, therefore

$$Y = v_{0V}\left(\dfrac{v_{0V}}{g}\right) - \dfrac{g}{2}\left(\dfrac{v_{0V}^2}{g^2}\right) = \dfrac{v_{0V}^2}{g} - \dfrac{g}{2}\dfrac{v_{0V}^2}{g^2} = \dfrac{v_{0V}^2}{g} - \dfrac{v_{0V}^2}{2g} = \dfrac{v_{0V}^2}{2g}.$$

There you have it. You have constructed a complete algebraic solution for the time of flight, maximum height, and range in terms of only the information given. Each year when I completed presenting this magnificent solution I would turn and ask my students, "Isn't this a thing of beauty?" I nearly always received the cynical resounding chorus, "Yeah, right!"

Notice that we took a very complicated problem and analyzed it into three major parts. You used the information discussed in this chapter to guide your thinking critically about each part until you arrived at a partial solution. You realized that the common item in the partial solutions could be found in terms of the information given by using the definition of acceleration. Substituting that information into all three you accomplished quite a feat. Synthesizing the results of your analysis gives a complete solution to the original problem. This process will serve you well in solving many problems, and I don't mean just physics problems. We are often stumped by a complex problem because we do not take the time to analyze it into its parts. Solve each part one by one until you can put all their solutions together to create a solution to the original problem.

Physics Phun

1. An Idaho potato is launched from a spud gun at an angle of $(3.00 \times 10)°$ with a muzzle velocity of $5.00 \times 10 \, \dfrac{m}{s}$. What is the time of flight, maximum height, and range of the tater?

2. A steel ball of mass 114 g is launched from ground level with an initial velocity of $615 \, \dfrac{m}{s}$ at an angle of $(4.50 \times 10)°$. What is the time required to reach maximum height? What is the acceleration of the ball at the point of maximum height? What is the velocity of the ball at the point of maximum height? What is the net force on the ball at the point of maximum height? How long does it take the ball to descend from the maximum height? Calculate the time of flight, the maximum height, and the range of the ball.

A Baseball from Center Field

I think everyone likes baseball. They like to play the game or watch it or both. I doubt if many people watch the game with the same awareness that I am going to suggest to you. Of course, if you have worked through all of the examples and Physics Phun problems, the applications of the physics principles here may be anticlimactic. It is good for you to see that the ideas that have been discussed in this chapter are applicable to everyday life, even to our favorite pastimes. Think of the baseball as a good approximation of an ideal projectile. Apply the analysis of a projectile in motion to the flight of a baseball on a cool San Francisco evening.

Suppose a batter hit a 145 g baseball into deep center field and the ball was caught. A runner at third base tags up and runs for home plate. The center fielder throws the ball to the catcher, it arrives 4.0 seconds after it was thrown, and the catcher catches the ball at the same height from which it was thrown. It would be great if we could figure out whether the runner is tagged out or not, but that depends on a lot of conditions we are not prepared to handle. We can find out what maximum height the ball reaches, what is the initial vertical component of velocity, what is the net force on the ball at maximum height, and what is the acceleration at maximum height.

Begin the solution with the time of 4.0 seconds. That is called the time of flight in our analysis of projectile motion, so $T = 4.0s$ and $t = 2.0s$ where t is the time required to reach maximum height. You solved for t algebraically and found $t = \dfrac{v_{0V}}{g}$, so $v_{0V} = gt$, and $v_{0V} = \left(980\,\dfrac{cm}{s^2}\right)(2.0s) = 2.0 \times 10^3\,\dfrac{cm}{s}$ observing two significant figures. The maximum height for a projectile was calculated to be $Y = \dfrac{v_{0V}^{\,2}}{2g}$, so

$$Y = \frac{\left(2.0 \times 10^3\,\dfrac{cm}{s}\right)}{2\left(980\,\dfrac{cm}{s^2}\right)} = 2.0 \times 10^3\,cm = 2.0 \times 10 m.$$

Remember the maximum height is measured from the level of launch. At maximum height the acceleration of the ball is the acceleration due to gravity and is the same as for all objects near the surface of the earth. The vertical velocity is zero at that point but not the acceleration. Finally, the net force on the ball at maximum height is the same as for any other place in the flight because the baseball is treated here as an ideal projectile. That means by Newton's second law, $W = mg = 145g \times 980\,\dfrac{cm}{s^2} = 1.42 \times 10^5\,dynes$.

There, I have ruined it for you. You will never look at a baseball traveling from center field to home plate in the same way again. Sorry about that. Physics has a way of removing a lot of mystery and exploding a lot of myths. For instance, did you know that there is no such thing as centrifugal force? I will guide you through a proof of that in the section "Force in Relation to Mass and Distance" in Chapter 8.

The Hang Time of a Punt

The example of a baseball as a projectile is not a unique situation. You can discuss the behavior of other sports toys as well. The path of a hockey puck could be interesting but a bit dangerous, I would think. A tennis ball or a Ping-Pong ball could be studied as long as it is not a windy day and there is not English (spin) on either. By the way, we will have to be careful with a baseball on a windy day, too, as well as a football.

Both can be good applications of the path of a projectile, and we will specify that we are making our study on a calm, cool evening. We take a look at football next, and even though we are focusing on the equipment used to play the game, the skill of the players is certainly a determining factor in all of our discussions.

I know that you have watched a football game during which you heard the announcer discussing the leg of the punter. That is, he has a strong leg if he can give the football a hang time of 4.0 seconds or more. The hang time of the football is the time of flight of a projectile. Suppose the hang time for a punt of a football with a mass of 0.028 slugs is 4.5 seconds. How many feet does it go in the air above the punter's toe? What is the weight of the football at maximum height? What is its initial vertical velocity?

You know that the weight of the football is the same everywhere near the earth's surface. Newton's second law enables you to calculate the weight as

$W = mg = 0.028 \, slugs \left(32 \frac{ft}{s^2} \right) = 0.90 \, lb$. Since the time to reach maximum height is

given by $t = \frac{v_{0V}}{g}$, then $v_{0V} = gt = 32 \frac{ft}{s^2} \times 2.25s = 72 \frac{ft}{s}$. The maximum height you

found to be given by $Y = \frac{v_{0V}^2}{2g}$, then $Y = \frac{\left(72 \frac{ft}{s} \right)^2}{2 \times 32 \frac{ft}{s^2}} = 81 ft$.

As you can see, I used the algebraic solutions we developed in this chapter. That is fine as long as you can develop all of those expressions on your own. If you have difficulty developing those ideas, solve problems of this type by beginning with the particular type of motion involved and solve each problem algebraically first. Once you do that a few times, you will find that the whole solution is second nature to you and you can visualize each step while associating the solution with the diagram of the path of the projectile.

The Least You Need to Know

- Describe the motion of a dense object when it is launched horizontally near the surface of the earth.

- Describe the motion of a dense object when it is dropped near the surface of the earth.

- Identify information needed to describe the motion of a projectile launched into the air at an angle with the horizontal.

- Analyze the motion of a projectile into two parts depending on the type of motion.

- Use the analysis of the motion of a projectile to determine the time of flight, maximum height, and range.

Chapter 8

Newton's Universal Law of Gravitation

In This Chapter

◆ Paths of masses due to force and distance

◆ Movements of planets

◆ The attraction of the earth for the sun

◆ How an earth satellite stays in orbit

◆ Use of a satellite to measure mass

Recently I read about a test that revealed that a large percentage of those who took the test did not know that the earth orbits the sun and that it takes about 365 days to make one complete orbit. This chapter will not only make you acutely aware of that relationship, but also that the earth makes one complete rotation about its axis in about 24 hours. This information can help you gain insight into the world around you and beyond.

The next time you watch the moon rise big and beautiful above the horizon, think about the motion that is involved. You will find that a lot of people who are interested in physics are dreamers with "stars in their eyes."

There is nothing wrong with enjoying the beauty of your world that some of those people have helped us to see and understand. Prepare to be amazed by a glimpse, in this chapter, of the work produced by the minds of some of those giants.

How Is Force Related to Mass and Distance?

In the previous chapter we discussed the path of a projectile in terms of the horizontal component and the vertical component of the instantaneous velocity. I drew a diagram of the path of the projectile but we never named the path, and it is helpful to identify curves with a name. Consider the two equations we used to describe the motion of the projectile. We used $x = v_{0H}\tau$ and $y = v_{0V}\tau - \frac{1}{2}g\tau^2$. These two equations are called parametric equations with the parameter τ.

Solving for the parameter in either equation and substituting the results in the other equation can eliminate the parameter. Since I am intellectually lazy, I choose to solve for the parameter in the first equation and find $\tau = \dfrac{x}{v_{0H}}$. Substitute for τ in the other equation and get $y = v_{0V}\left(\dfrac{x}{v_{0H}}\right) - \dfrac{1}{2}g\left(\dfrac{x}{v_{0H}}\right)^2 = \dfrac{v_{0V}}{v_{0H}}x - \dfrac{1}{2}g\left(\dfrac{x}{v_{0H}}\right)^2 = \dfrac{v_{0V}}{v_{0H}}x - \dfrac{g}{2v_{0H}^2}x^2$.

The last equation is the equation of a parabola. We have a name for the nice smooth curve that I referred to in the last chapter as the path of the projectile: it is a parabola. Is that a big deal or what? Watch carefully because we are headed toward a collision with a physics myth that I promised you earlier.

The force of gravity is the net force on the projectile we discussed earlier. The path of the projectile is a parabola as shown in Figure 8.1.

Figure 8.1

The components of the force of gravity on a projectile influence the motion of the projectile.

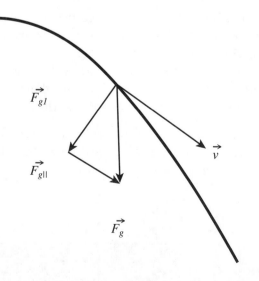

The force of gravity is shown acting on the object at one point along the path. The components of the force of gravity perpendicular to instantaneous velocity, $\vec{F}_{g\perp}$, and parallel to the instantaneous velocity, $\vec{F}_{g\parallel}$, at that point are shown as well. You know by Newton's first law that an object tends to continue moving in a straight line unless acted upon by a net force.

The object traveling the parabolic path would continue traveling tangent to the path at any point if it did not have $\vec{F}_{g\perp}$ deflecting it out of the straight-line path at every instant that it is in the air. The component $\vec{F}_{g\parallel}$, on the other hand, acts parallel to the instantaneous velocity causing the object to speed up along the path. Therefore, the force of gravity on the object has one component deflecting the object into a curved path and another component speeding the object up along the path creating the nice smooth parabolic *trajectory* in Figure 8.1.

Please take a moment to go back to Chapter 4 and review briefly the derivation of the equation for centripetal acceleration in the section on "Centripetal Acceleration."

The derivation yields the equation $a_c = \dfrac{v^2}{R}$, which is the magnitude of the centripetal acceleration of an object traveling in a circular path at a constant speed. Another way of expressing the same idea is with the vector equation $\vec{a}_c = -\dfrac{4\pi^2 \vec{R}}{T^2}$ where $\vec{R}$ is the positive position vector from the center to the object at any time and T is the period or the time required for the object to complete one revolution.

Plain English

The **trajectory** is the arced or curved path of a projectile. **Centripetal force** is the center-seeking force on an object at every instant and deflects the object into a circular path at a constant speed.

The centripetal acceleration is the only acceleration the object experiences when it is traveling in a circular path at a constant speed. Unlike the projectile in the parabolic path, there is no increase in the speed. The centripetal acceleration is always perpendicular to the instantaneous velocity and that means it is always pointed directly toward the center of the circular path. You know that the only way you expect acceleration is that, by Newton's second law, a net force is acting in the direction of the acceleration.

The magnitude of the force causing centripetal acceleration, by Newton's second law, is given by $f = ma = \dfrac{mv^2}{R} = \dfrac{m4\pi^2 R}{T^2}$, and is always directed radially inward toward the center of the circular path. For that reason, the force is called centripetal *force*. It is the only force causing acceleration. Therefore, centrifugal force does not exist!

Newton's Figs _____

Some people will tell you that centrifugal force does exist as sure as two plus two is four. You can respond that two plus two may not be four because you can show when discussing vectors that two plus two may be zero. Remember, the same people perhaps could care less what a vector is, so be nice.

If another force did exist, a centrifugal force, the object would accelerate in the direction of that force when the centripetal force is removed. That can be accomplished by cutting the string or rope or whatever is holding the object as it travels in a circular path at a constant speed. You find when that is done the object travels in a straight line at a constant speed, the speed it had when it was released. The direction it travels is tangent to the circular path at the instant it is released. There is nothing like some good old physics experiments to convince yourself.

Again I just fulfilled some promises to you. I helped you to calculate centripetal force and at the same time showed you that centrifugal force does not exist! Why were you told about centrifugal force all your life? It may be a misapplication of Newton's third law. You know that the correct application of that law to this situation is that the string pulls on the object and the object pulls on the string with an equal and opposite force. There is nothing pulling on the object radially outward from the center of the circular path.

Another situation you have probably experienced could easily give rise to the confusion. Suppose you are on the passenger side of the front seat when the driver successfully negotiates a sharp left hand curve. It seems to you that something is pulling you out of the passenger side of the car because of the tension you experience from the seatbelt. The problem here is you are in an accelerated frame of reference, the car, and Newton's laws do not apply to an accelerated frame of reference.

Newton's laws hold in *inertial frames of reference*, frames of reference either at rest or moving at a constant speed. An observer standing on the earth, an inertial frame of reference, would explain to you that nothing is pulling you out of the car. The seatbelt is deflecting your body into the circular path at every instant, keeping you from traveling in a straight line at a constant speed tangent to the path.

A type of motion that can be described at this time is *simple harmonic motion* because we know that centripetal force causes an object to move in a circular path at a constant speed.

Imagine that an object of mass m moves with a constant speed around a vertical circular path of radius R as shown in Figure 8.2. As m moves around the circle, the projection of m moves back and forth along the diameter, AB, of the circle. When m starts at A and makes one complete trip around the circle, the projection of m starts at A, then travels to B and back to A. All of this movement takes place in the period of motion of m, the time to make one complete revolution. The *period* of the motion of the projection of m is exactly the same as the period of the motion of m.

Physics Phun

An **inertial frame of reference** is a frame of reference that is at rest or moving at a constant speed. For our purposes, the earth is an inertial frame of reference. **Simple harmonic motion** is the to-and-fro motion caused by a restoring force that is directly proportional to the magnitude of the displacement and has a direction opposite the displacement. An example is the motion exhibited by a vibrating string or a simple pendulum. The **period** of simple harmonic motion is the time for the object in motion to complete one cycle. The path from A to B and back to A is one cycle for the projection of m, as in Figure 8.2.

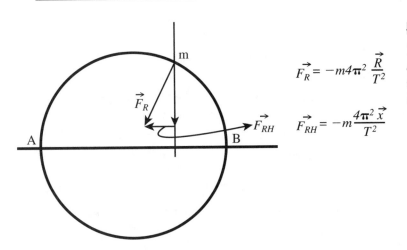

$$\vec{F_R} = -m4\pi^2\,\frac{\vec{R}}{T^2}$$

$$\vec{F_{RH}} = -m\,\frac{4\pi^2\,\vec{x}}{T^2}$$

Figure 8.2

The horizontal component of centripetal force models the restoring force for simple harmonic motion.

You know that the force on m is the centripetal force labeled $\vec{F_R}$ in Figure 8.2. You can think of the horizontal component of the centripetal force labeled $\vec{F_{RH}}$ in that diagram as the force acting on the projection of m. That is, this is a model related to uniform circular motion that can be used to explain another type of motion and its causes. Notice that as m moves counterclockwise from its first position in that diagram, the direction of $\vec{F_{RH}}$ is to the left until m gets to a point directly above the center of the circle. At that instant, $F_{RH} = 0$. As m continues counterclockwise, the direction of $\vec{F_{RH}}$ is now toward the right. It keeps that direction until m is at a point directly beneath the center of the circle, where $\vec{F_{RH}}$ becomes zero again. The direction of $\vec{F_{RH}}$ is toward the left after that point, and remains pointing left through the starting point and until m is at a point directly above the center. At that point F_{RH} becomes zero again and the motion repeats. Not only does the direction of $\vec{F}_{RH}$ change as m moves around the circle, but its magnitude changes also. When m is directly above the center, $\vec{F}_{RH} = 0$; it then increases in magnitude until m is at A, where $\vec{F}_{RH}$ is maximum. As m continues to move counterclockwise, $\vec{F}_{RH}$ decreases in magnitude and becomes zero when m is directly below the center of the circle. It then increases in magnitude and reaches maximum magnitude when m is at B.

The motion of the projection is said to be periodic with period T. That is, $\vec{F}_{RH}$ goes from zero to maximum to zero to maximum then back to zero while m is making one complete revolution. The complete trip of m in one revolution and the corresponding trip of its projection is called a *cycle*. The units of T can then be thought of as $\dfrac{s}{cycle}$. You are probably more familiar with the reciprocal of the period that is called the *frequency*. That is, the frequency is given by $f = \dfrac{1}{T}$ and the units of frequency can be thought of as $\dfrac{cycles}{s}$.

Plain English

A **cycle** is one complete trip for an object moving in a circular path at a constant speed as well as the corresponding trip of its projection on the diameter shown in Figure 8.2. The time for one cycle is called the period T, which means that you can think of the units of T as $\dfrac{s}{cycle}$. The **frequency** of the motion is the reciprocal of the period and is given by $f = \dfrac{1}{T}$; it has units of $\dfrac{cycles}{s}$ or just $\dfrac{1}{s} = s^{-1}$ since a cycle is not a unit of measurement. Another common unit of measurement of frequency is the Hertz, where $1\ Hz = 1s^{-1}$.

Just as the centripetal force is directed opposite the radius vector of the circle, the horizontal component of the centripetal force is directed opposite the displacement of the projection along the diameter of the circle. The displacement is labeled x and is measured from the center of the circle to the projection.

Notice that $\vec{F}_{RH}$ is maximum when the displacement is maximum and $\vec{F}_{RH} = 0$ when the displacement is zero. While m travels at a constant speed, the projection changes velocity. The velocity of the projection is a maximum when the displacement is zero; in fact, at that instant its velocity is the same as the instantaneous velocity of m. The velocity of the projection is zero when the displacement has a maximum value. All of this information is necessary as when identifying the motion of the projection of m as simple harmonic motion.

Simple Harmonic Motion in a Spring

If you hang a mass from a spring, the mass will stretch the spring a certain amount and then come to rest. You know that situation is established when the pull of the spring upward on the mass is equal to the pull of the force of gravity downward on the mass. The system, spring and mass, is said to be in *equilibrium* when that condition is met.

If you displace the mass up or down from the equilibrium position and release it, the spring will undergo simple harmonic motion caused by a force acting to restore the vibrating mass back to the equilibrium position. That force, called the restoring force, is directly proportional to magnitude of the displacement and is directed opposite the displacement. The necessary condition for simple harmonic motion is that a restoring force exists that meets the conditions stated symbolically as $\vec{F_r} = -k\,\vec{x}$, where k is the constant of proportionality and $\vec{x}$ is the displacement from the equilibrium position. The minus sign, as usual, indicates that $\vec{F_r}$ has a direction opposite that of $\vec{x}$.

Plain English

A system is in **equilibrium** when the sum of the forces is zero. If one force is used to balance the effects of two or more other forces, then that force is called the equilibrant and is equal in magnitude and opposite in direction to the resultant of the other forces.

You can make sense of all of these pieces of information by stating $\vec{F_r} = \vec{F_{RH}}$. That means that the restoring force causing simple harmonic motion is exactly the same as the horizontal component of the centripetal force that causes the motion of the projection along the diameter in Figure 8.2. Considerable information about simple harmonic motion can be gained from that statement:

$\vec{F_r} = \vec{F_{RH}}$ The restoring force is equal to the horizontal component of the centripetal force.

$-k\,\vec{x} = -m\dfrac{4\pi^2 \,\vec{x}}{T^2}$ Substituting definitions of both.

$k = \dfrac{m4\pi^2}{T^2}$ Since $-\vec{x}$ are the same in both cases, the constant k is the same in both cases.

$T^2 = \dfrac{m4\pi^2}{k}$ Multiplying both sides of the equation by $\dfrac{T^2}{k}$.

$T = 2\pi\sqrt{\dfrac{m}{k}}$ Regrouping and finding the principal square root of each member of the equation.

The last equation states that the period of simple harmonic motion is calculated using the mass of the object and the constant of proportionality for the displaced object. It reveals that the period is directly related to the mass and inversely to the constant of proportionality. Applying the idea to the spring, for a given spring, the larger the mass the longer the period of vibration. Also, given a certain mass, the stiffer the spring, the shorter the period of vibration. If you think about these statements, both make sense from similar experiences of yours. Better yet, you can do your own experiment and check the validity of these statements.

Simple Harmonic Motion and the Simple Pendulum

You know about the *simple pendulum*. When set in motion using small angles, the simple pendulum undergoes simple harmonic motion. Small angles here are angles less than 10°. The reason is because if you use a large angle, say 90°, and release the pendulum, it will tend to drop straight down and bang around until there is a smooth swing. You want a nice smooth swing from the beginning in order to have simple harmonic motion.

Look at the simple pendulum in Figure 8.3 and see how the derivations of the period of simple harmonic motion applies to this familiar object undergoing simple harmonic motion. You know that it is simple harmonic motion when you show that there is a restoring force directly proportional to the displacement from the equilibrium position and directed opposite the displacement.

Plain English

A **simple pendulum** is a physical object made up of a mass suspended by a string, rope, or cable from a fixed support. It is called a simple pendulum because the string, rope, or cable has a negligible amount of mass compared to the mass of the object being supported. A physical pendulum is a physical object having the mass distributed along the full length of the pendulum and not having most of the mass concentrated in one place as in the simple pendulum.

Notice that the pendulum is made up of a string of length l supporting a mass m. The vertical position of the pendulum is the equilibrium position. The force of gravity on the mass m is vertically downward, of course, and the string pulls upward with a tension equal to the weight of the mass. I did not draw the force of gravity in that position to avoid a confusing diagram. The pendulum was displaced to the right by angle θ, which should be less than 10°. I had to draw the angle large enough so that you can see the geometrical relationships in the diagram.

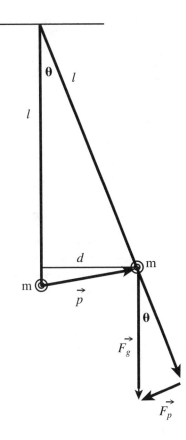

The force of gravity was drawn at the second position of the pendulum and then is resolved into two components. One component is parallel to the string (keeping the tension in the string) and the other component, labeled $\vec{F}_p$, is perpendicular to the string. The acute angles in the two right triangles are labeled θ to indicate that they are equal. Two angles are equal if they have their sides parallel right side to right side and left side to left side. That means the two triangles are similar. Two right triangles are similar if an acute angle of one is equal to an acute angle of the other. Therefore:

$\dfrac{F_p}{d} = \dfrac{F_g}{l}$ Corresponding parts of similar triangles are proportional.

$F_p = \dfrac{d}{l} F_g$ Multiplying both equations by d.

$F_p = \dfrac{d}{l} mg$ By substitution.

As *m* swings to and fro, its path is a small arc of a circle. For small angles, the chord *p* is about equal in length to the corresponding arc, thus the chord *p* is about equal to the semi-chord d. The vector $\vec{p}$ is the displacement of m from the equilibrium position. Therefore, $\vec{F_g} = -\dfrac{mg}{l}\vec{p}$, substituting p for d and since $\vec{p}$ and $\vec{F_p}$ are in opposite directions we use a minus sign. This shows that there is a restoring force that is directly proportional to the displacement and acts in a direction opposite the displacement.

That means that $k = \dfrac{mg}{l}$ and since for simple harmonic motion $T = 2\pi\sqrt{\dfrac{m}{k}}$ we can substitute for k and find $T = 2\pi\sqrt{\dfrac{ml}{mg}} = 2\pi\sqrt{\dfrac{l}{g}}$. Therefore, the simple pendulum meets the criteria for simple harmonic motion and has a period, $T = 2\pi\sqrt{\dfrac{l}{g}}$. Among other things, the simple pendulum can be used to measure the acceleration due to gravity.

You may find some of these ideas challenging and interesting. As a young man, Galileo was challenged by this phenomenon and observed a swinging lamp in the sanctuary of the cathedral at Pisa. He used his pulse to time the cycles of motion of the lamp and found that cycles were completed in the same time no matter what the *amplitude* of the motion of the lamp, assuming small angles of arc. You may find simple experiments like his very enlightening. Maybe you would not do a science experiment at church or maybe you think there is no better place. After all, a science laboratory is not required to do interesting physics experiments.

Plain English

The **amplitude** of simple harmonic motion is the maximum displacement of the motion.

Physics Phun

1. Calculate the period of a simple pendulum having a length of 25 cm. The pendulum is near the surface of the earth.

2. A simple pendulum having a length of 55.8 cm is used to measure the acceleration due to gravity. What is the acceleration due to gravity if the period of the pendulum is 1.50 s?

3. How much centripetal force is required to keep a 2.0 g object moving in a circular path of radius 111 cm with a period of 0.61 s?

4. What is the frequency of the pendulum in problem (2)?

5. Discuss the acceleration of an object that is undergoing simple harmonic motion. What can you say about its acceleration when it has maximum velocity?

You have found that centripetal force varies inversely with the distance and directly as mass. The restoring force for simple harmonic motion varies directly as the distance. What do you expect for the force of gravity at some distance from the earth? A good way to find an answer is to consider an object not near the surface of the earth but at a considerable known distance from the earth. Canada launched a satellite, Alouette I, into an orbit that was nearly circular. Any satellite must be given a velocity that is large enough to keep it in orbit. If the orbit is nearly circular, the force keeping the satellite in orbit is centripetal force. The centripetal force must be the force of gravitational attraction of the earth for the satellite at that altitude. You can use what you know about centripetal force to discuss the motion of a satellite orbiting the earth as well the cause of that motion.

The satellite has an altitude above the surface of the earth of $h = 1.01 \times 10^6 m$ and the radius of the earth is $R = 6.38 \times 10^6 m$. The mass of the satellite at that altitude is the same as its mass on earth. That means that the acceleration due to gravity must be different at that altitude than it is near the surface of the earth. By the time you complete this chapter you will be able to calculate the acceleration at that altitude, but for now I state that the acceleration due to gravity at $h = 1.01 \times 10^6 m$ above the surface of the earth is $g_a = 7.30 \ m/s^2$. What is the velocity that the satellite must have to maintain that altitude? Use the analysis suggested here to calculate a solution. Remember this is just one approach to a solution.

$f_c = f_{ga}$ The centripetal force is equal to the force of gravity at that altitude.

$m\dfrac{v^2}{R_a} = mg_a$ Substituting definitions of the centripetal force at that altitude is equal to force of gravity at that altitude.

$\dfrac{v^2}{R_a} = g_a$ Dividing both equations by m, which is the same at both positions.

$v^2 = R_a g_a$ Multiplying both equations by R_a.

$v = \sqrt{(R+h)g_a}$ Solving for v and substituting for R_a.

$v = \sqrt{\left[\left(6.38 \times 10^6 m + 1.01 \times 10^6 m\right)\left(7.30 m/s^2\right)\right]}$ Substituting numerical values.

$v = 7.34 \times 10^3 m/s$

Make the answer a little more understandable by expressing the velocity in miles/hr. You will need the defining equation 1 mi = 1610 m in order to make the calculation:
$v = \left(7.34 \times 10^3 m/s\right)\left(1/1610 \ mi/m\right)\left(60s/\min\right)\left(60 \min/hr\right) = 16,400 mi/hr.$

Another interesting detail that we can gain from this example is a comparison of the force of gravity at $1.01 \times 10^6 m$ with the force of gravity at the surface of the earth: $\frac{f_{ga}}{f_{g0}} = \frac{mg_a}{mg_0} = \frac{g_a}{g_0}$.

The ratio of the force of gravity at the altitude of the satellite to the force of gravity at the surface of the earth (h=0) is equal to the ratio of the acceleration due to gravity at the altitude to the acceleration due to gravity at the surface of the earth.

Using values in this example, $\frac{f_{ga}}{f_{g0}} = \frac{7.30m/s^2}{9.80m/s^2} = 0.745$. That is, about 74.5% of the force of gravity at the surface of the earth is the force of gravity on the satellite at $1.01 \times 10^6 m$ above the surface of the earth. The force of gravity appears to vary inversely with the distance, but where does mass enter this discussion? There were two masses involved here, the mass of the earth and the mass of the satellite, and we used the mass of the satellite alone. There is more to this story and you are seeing it unfold.

The Motion of Heavenly Bodies

There is a long history of the efforts of men to explain the motion of heavenly bodies. Many models were contrived to explain the strange motion of the planets. I share some of the work of a few scientists of more recent times. Tycho Brahe replaced the system devised by Copernicus with his own geometric system where the sun goes around the earth and the other planets go around the sun. He was a skilled experimenter who made careful observations and recorded data accurately identifying the positions of planets and fixed stars.

Johannes Kepler worked in Tycho's laboratory during the last year or so of Tycho's life. Kepler had great respect for the work of Tycho and his records that must have filled volumes of books. Kepler felt that there must be an easier way to describe the motion of heavenly bodies than to record their positions in tables on page after page of laboratory records. Kepler was not a skilled experimenter like Tycho but he had a strong mathematics background. Mathematics was not one of Tycho's strengths. Kepler had that special scientific sense that all of the volumes of Tycho's observations could be replaced by a simpler idea. Simplicity is a goal of all scientists when they are seeking an explanation for physical phenomena.

Kepler took the work of Tycho Brahe and developed a mathematical model to replace all of the observations Tycho had so carefully recorded. But Kepler had so much respect for the work of Tycho Brahe that he questioned one of his own mathematical models when it disagreed with an observation by Tycho by a fraction of a degree of arc. He abandoned that model based on eccentric circles for the paths of the planets because he did not think that Tycho Brahe could make such a mistake in his observations.

Eventually his hard work came to fruition when he developed a mathematical model that not only agreed with the work of Tycho but also made predictions that could be verified by observation. His model is summarized in the statement of Kepler's three laws listed here:

1. The planets travel in elliptical paths with the sun at one focus.

2. A line joining the sun and a planet sweeps out equal areas in equal times.

3. The ratio of the cube of the mean radius to the square of the period of revolution is the same for all planets. A simple statement of Kepler's third law is
    $k = \dfrac{R^3}{T^2}$.

Gravitational Force Between any Two Bodies

Kepler's laws describe the motions of the planets. Newton used the work of Kepler to discover the causes of the motions of the planets. He used the fact that even though the paths of the planets were elliptical they are so nearly circular that circular paths were used to calculate the force causing the motion. He used his own idea that

$$F_{sp} = m_p a_p = m_p \frac{4\pi^2 R}{T^2}.$$

Remember that the equations state in words the force of attraction of the sun for a planet is equal to the product of the mass of the planet and the centripetal acceleration. Another way of stating the same thing is the attraction of the sun for a planet is equal to the mass of the planet divided by the square of the period of revolution of the planet around the sun times the mean radius of the path of the planet times a constant. Newton thought that there would be an inverse square relationship for the distance between the planet and the sun. He used Kepler's third law to express the force in terms of R and m only.

That is, since $k = \dfrac{R^3}{T^2}$ then $T^2 = \dfrac{R^3}{k}$ and the force can be expressed as

$F_{sp} = m_p \dfrac{4\pi^2 R}{\dfrac{R^3}{k}} = m_p \dfrac{4\pi^2 k}{R^2}$. He used the same idea to calculate the force of attraction

of the earth for the moon and arrived at a similar expression except the constant k was

associated with the earth, that is, $F_{em} = m_m \dfrac{4\pi^2 k_e}{R^2}$. He felt that there must be some

property of the earth involved in the constant k_e so the force of the sun on a planet

must have a similar relationship, that is, $F_{sp} = m_p \dfrac{4\pi^2 k_s}{R^2}$.

Satisfied he had found that the force was directly proportional to the mass of the planet and inversely proportional to the square of the distance to the planet he wondered what the constants $4\pi^2 k_e$ and $4\pi^2 k_s$ and similar expressions for other planets meant. He inferred that it must involve some property of the body about which the object is orbiting. He thought that property is the mass of the body and stated $4\pi^2 k_s \propto m_s$ so $4\pi^2 k_s = Gm$ and similarly $4\pi^2 k_e = Gm$. Then the force can be stated $F_{se} = \dfrac{m_e G m_s}{R^2}$, and by his third law if the sun attracts the earth then the earth must attract the sun with the force $F_{es} = \dfrac{m_s G m_e}{R^2}$. When these are simplified they are one and the same, that is, $F_{es} = F_{se} = \dfrac{G m_e m_s}{R^2}$ and the directions are opposite. He then generalized this result to state his universal law of gravitation: Any two particles in the universe attract each other with a force that is directly proportional to the product of their masses and inversely proportional to the distance between their centers of mass. Cavendish determined an accurate value of G experimentally and recently more accurate values of G have been determined as technology for making such measurements has improved. The value of G is 6.67×10^{-11} and the units were assigned in the same way we assigned units before for the constant in Newton's second law. The universal gravitational constant, G, is $6.67 \times 10^{-11} \dfrac{N - m^2}{kg^2}$. Newton's universal law of gravitation is stated symbolically as $F = \dfrac{G m_1 m_2}{R^2}$.

Of course, no one really knows how Newton arrived at his Universal Law of Gravitation. I have tried to give you an idea of how he might have reasoned through this work. I mainly want you to be aware of how great ideas evolve as a result of the scientific attitude discussed earlier. Kepler started with Brahe's exhaustive records of observations and replaced them with a simple set of mathematical laws. Newton built on the work of Galileo and Kepler to develop his ideas. Newton's Universal Law of Gravitation explains the motions of planets as well as causes for the motions. It had one difficulty with an explanation for the behavior of the orbit of Mercury. It took Einstein's theory of gravity to explain that little detail. Do you see how scientists do not depend on absolute truth? Instead of believing that Kepler's work was the last word in the description of the motion of heavenly bodies, Newton developed his theories that enable us to understand things that were impossible before he gave us his work. Even that great work was not absolutely true. Einstein built on Newton's work and who knows, you may be next!

The Acceleration of Satellites of the Earth

I gave you the acceleration of gravity at a high altitude in an earlier problem and told you that you will be able to calculate the acceleration due to gravity at any altitude above the earth. Before we do that, calculate the weight of a body at any altitude above the earth. Using Newton's Universal Law of Gravitation will enable you to do that. Let m_e be the mass of the earth and m_b be the mass of a body and let R_e be the radius of the earth and R_h the distance from the center of the earth to the body above the earth. An easy way to compare two things is to find the ratio of the two:

$$\frac{W_e}{W_h} = \frac{\dfrac{Gm_e m_b}{R_e^2}}{\dfrac{Gm_e m_b}{R_h^2}} = \frac{\dfrac{1}{R_e^2}}{\dfrac{1}{R_h^2}} = \frac{R_h^2}{R_e^2} = \left(\frac{R_h}{R_e}\right)^2 \text{ and that means } W_h = \left(\frac{R_e}{R_h}\right)^2 W_e.$$

Suppose that a 160 lb man at the surface of the earth is placed at an altitude 4000 mi above the earth. What will be his weight at that altitude? The radius of the earth is about 4000 mi. Substituting numerical values in the algebraic solution to this problem we get $W_h = \left(\dfrac{4000mi}{4000mi + 4000mi}\right)^2 \times 160lb = \left(\dfrac{1}{4}\right) \times 160lb = 40lb.$ A nice way to lose weight but you would have a whole lot of trouble breathing at that altitude. This is another example of the beauty of algebra in the solution of a problem. Had we used numerical values at the beginning, we would have a mess that makes it difficult to check the work. Here you have a complete solution that can be read like an essay.

A Johnnie's Alert

Always include the radius of the earth as part of the distance to a satellite when it is orbiting at a given altitude. It is easy to forget that the acceleration or weight at the altitude depends on the distance made up of the altitude plus the radius of the earth. The distance in Newton's equation is the distance from the center of a mass to the center of the other mass.

Suppose you know that a satellite is at a known altitude above the earth, and for some reason you need to know the acceleration of the satellite. You can use Newton's second law and the Universal Law of Gravitation to calculate the acceleration due to gravity at any altitude.

$$w_{sh} = f_{gsh}$$

$$m_s g_{sh} = \frac{Gm_s m_e}{R_h^{~2}}$$

$$g_{sh} = \frac{Gm_e}{R_h^{~2}}$$

The algebraic solution states: The acceleration due to gravity at any altitude is equal to the universal gravitational constant times the mass of the earth divided by the square of the distance between the center of the earth and the satellite. Calculate the acceleration due to gravity at the altitude used earlier in the example, $h = 1.01 \times 10^6 m$. The mass of the earth is $5.98 \times 10^{24} kg$.

$$g_{sh} = \frac{\left(6.67 \times 10^{-11} \frac{N-m^2}{kg^2}\right) \times \left(5.98 \times 10^{24} kg\right)}{\left(1.01 \times 10^6 m + 6.38 \times 10^6 m\right)} = 7.3 \frac{N}{kg} = 7.3 \frac{\frac{kg-m}{s^2}}{kg} = 7.3 \frac{m}{s^2}$$

You just calculated the value I gave you to work with earlier. Any time you need the acceleration due to gravity at any altitude, you can calculate it.

Calculate the centripetal acceleration of a natural satellite of the earth. The moon is about $3.8 \times 10^8 m$ from the earth.

$$a_c = g_{mh} = \frac{\left(6.67 \times 10^{-11} \frac{N-m^2}{kg^2}\right) \times \left(5.98 \times 10^{24} kg\right)}{\left(3.8 \times 10^8 m\right)^2} = 2.8 \times 10^{-3} m/s^2$$

The mass of the moon is $7.34 \times 10^{22} kg$. What is the centripetal force needed to keep the moon in orbit?

$$f_c = f_g = m_m g_{mh} = \left(2.8 \times 10^{22} kg\right)\left(2.8 \times 10^{-3} m/s^2\right) = 7.84 \times 10^{19} N$$

Measuring the Mass of the Earth

Did you ever wonder how they did that? How they ever measured the mass of the earth? You know with just a bit more information you can use the results of modern technology to calculate the mass of the earth. Scientists study the mass of distant heavenly bodies using the same technique that is outlined in this section. First, the bit of information you need is the period of the satellite used in our earlier example. Alouette I completes an orbit every 105.4 minutes; that is, the period of the satellite is 105.4 minutes. You know that a centripetal force equal to the force of gravity keeps the satellite in orbit. Use that information along with ideas developed in this chapter to construct a solution to the calculation of the mass of the earth.

$$f_c = f_g$$

$$m_s \frac{4\pi^2 R}{T^2} = G\frac{m_e m_s}{R^2}$$

$$\frac{4\pi^2 R}{T^2} = G\frac{m_e}{R^2}$$

$$m_e = \frac{4\pi^2 R^3}{GT^2}$$

$$m_e = \frac{4\pi^2 \left(6.38\times10^6\,m + 1.01\times10^6\,m\right)^3}{\left(6.67\times10^{-11}\,\frac{N-m^2}{kg^2}\right)\left(105.4\,min\times 60\,\frac{s}{min}\right)^2}$$

$$m_e = 5.97\times10^{24}\,kg$$

Notice that I gave you the major steps in the solution and left the reasons for each step along with calculations for you to do. Make sure you check my work because you know that I made a mistake once before but fortunately you caught the error. It can happen again, not intentionally but out of my ignorance.

Physics Phun

1. Calculate the acceleration due to gravity at an altitude of $5\times10^5\,m$ above the earth.
2. Calculate the weight of a girl at an altitude of $6.6\times10^5\,m$ if her mass is 55 kg.
3. Calculate the mass of the earth if an orbiting satellite has a period of 100 min and an altitude of $7.5\times10^5\,m$.

The Least You Need to Know

- Identify Kepler's three laws.
- Calculate the acceleration due to gravity at any altitude.
- Explain how to use a simple pendulum to measure the acceleration due to gravity.
- Calculate the centripetal force on an object moving in a circular path at a constant speed.
- Use the horizontal component of the centripetal force to discuss simple harmonic motion.

Part 3

What Happens When Large Things Move?

So far you have spent most of your time learning to describe motion and dealing with the causes of motion. Now you are about to use all of that information to look at objects while they are in motion or before they have started to move. Trains, cars, billiard balls, and mud balls are some of the objects we can consider as we look at objects in motion. They may collide. Some may stick to the wall. Some moving objects may stick together, or collide without ever touching each other. Exploring the physics involved in situations such as these will be your challenge for a while.

Chapter 9

Work and Power

In This Chapter

- ◆ Me work?
- ◆ Calculating work
- ◆ The easier the better
- ◆ You need power

Has anyone ever told you that they are working themselves to death? Their concern may be true in one sense of the word, but you will find that scientists have a much different thing in mind when they talk about work. Someone may tell you that they use a certain breakfast bar to give them more power to start the day. Believe that person or not, you may have a different idea about the meaning of that claim when you see what power means to a scientist. You will also see that you do not buy power from a power company. More than likely the number of horses under the hood of your car will take on an altogether different meaning. Read this chapter with an open mind and prepare to cause a myth or two to evaporate and become part of the genie smoke.

Do You Work?

You might think that is a very silly question. Everybody knows that everyone works—or almost everyone. There is a lot of talk at this time about many people being out of work. You may mean that you work if you sit at a desk and write like crazy trying to meet a deadline. You may mean that you work hard at studying or maybe even work hard at playing. My son told me once that it is really hard work to grasp the monkey bars and support the body for any length of time. It may be true that in many of these instances work was done, but what you mean by work is not the work that was done.

Work is a very important concept in science. Physics explores the concept of work in detail by explaining what is meant by work, how work can be calculated, and by what units it is measured. You will probably become so wrapped up in these ideas that you will talk about it with your fellow workers at work! It is my intent to help you to at least think about a whistle while you work.

Imagine that you grasp the monkey bars with your hands and support your body like my son did. Shortly your arms tremble from supporting your weight. You probably would challenge me to tell you that you are not working. I accept the challenge, and say that I will use the scientific meaning of work to face the challenge. In a scientific sense, you will not do work on the monkey bars until you lift yourself a certain distance. The reason is that you must do two things in order to accomplish work in a scientific sense. A force must be applied to an object, and the object must be displaced in the direction of some component of that force. Some might think that work is just the product of force and displacement. You must be careful and state exactly what is meant by that idea because it can be very misleading. I won't leave you dangling on the monkey bars for long, but I do want you to be aware of some pitfalls when you tackle this new notion of work.

We will work our way through this mess soon, but first make note of some very special properties of the elements of work. We have used the idea of force and displacement. Both of these quantities are vector quantities. Work is a scalar quantity. When you take more mathematics, you will know how that works out. I state it now for you to keep in mind as we pursue an introduction to work. Since work is a scalar quantity, it has no direction associated with it. However, in order to understand what we mean by work in a scientific sense, we must deal with the vector quantity of force as well as displacement. It sounds confusing but it will become clear with some examples. We must be careful and not jump to conclusions or else we could jump on board with the person who thinks that work is just force times distance as sure as two plus two is four! It is a good thing that you will not fall for that one!

How Much Work?

That is a question that we can answer in this section. You know that work involves displacement and force and that work is a scalar quantity. You can calculate work by multiplying the magnitude of the applied force in the direction of the displacement of an object by the magnitude of the displacement. You may find that the force is not applied in the direction of the displacement. That is the pitfall we are to avoid. You need to know that if a force is applied to an object, and the object is displaced in a direction different from the direction of the force, then you must use the component of the applied force that acts in the direction of the displacement, Δx. This is shown in Figure 9.1 and Figure 9.2.

The work done to move the object along a horizontal surface in Figure 9.1 may be calculated simply as $W_k = F_A \Delta x$. This is a case where the force and displacement are in the same direction so the calculation is straightforward.

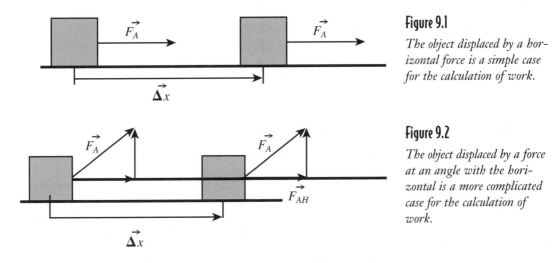

Figure 9.1

The object displaced by a horizontal force is a simple case for the calculation of work.

Figure 9.2

The object displaced by a force at an angle with the horizontal is a more complicated case for the calculation of work.

The case diagrammed in Figure 9.2 is a little more complicated because as you can see the object does not move in the direction of the applied force $\vec{F_A}$. That means that we must resolve the force into components perpendicular and parallel to the displacement and use the component parallel to the displacement to calculate the amount of work done. The horizontal component of the applied force in Figure 9.2 is labeled $\vec{F_{AH}}$. The work done in that case is $W_k = F_{AH} \Delta x$. Notice what all of the quantities used in the expression for work are. However, you know that work is a scalar quantity, and that it may be calculated by multiplying the magnitude of the component of the applied force parallel to the displacement by the magnitude of the displacement.

Your next task is to consider the units involved when you calculate the work done. Using the expression $W_k = F_{AH}\Delta x$ we will determine the units in each system. The MKS system provides the N and m for units of force and displacement respectively, so the units for work are $W_k = $ (N)(m) or N-m. You know that work is a derived quantity so there is no surprise that the unit of work is newton times meter or N-m. The unit of 1 N-m is defined to be 1 joule, so you have a new defining equation to add to your list: 1 joule = 1 N-m. Using Newton's second law we know that $1\,N = \dfrac{1\,kg-m}{s^2}$, so we can express 1 joule in fundamental units as follows: 1 joule = 1 N-m = $1\dfrac{kg-m^2}{s^2}$.

There are problems that will require you to recognize the units of work in any one of the three forms. The unit of work in the CGS system turns out to be (dynes)(cm) or dyne-cm by using the same algebraic expression as a pattern. The unit of work in the CGS system is the erg, which is the root of the word *energy*. The relationships among all of these units then are expressed as 1 erg = 1 dyne-cm = $1\dfrac{g-cm^2}{s^2}$.

If you solve crossword puzzles, you probably recognize the erg as the unit of work in the CGS system. Finally, the unit of work in the FPS system, weird as always, is the ft-lb. There is no one unit of work in the FPS other than 1 ft-lb. The units of work expressed in fundamental units are $1\,ft\text{-}lb = \dfrac{1slug-ft^2}{s^2}$. You can now calculate work, and then express your answer in correct units. I know you are ready to have some physics phun!

Example 1: Calculating Work Using a Horizontal Force

Suppose that a force of 54 N is required to roll a 981 N safe 11 m across the floor. How much work is done? The answer to this problem requires some thought. The way the problem is worded, it appears that the force is applied in the direction the safe is moving and that 54 N is just enough force to move the safe at a constant speed. Furthermore, the safe stops when the force is removed. That means that the work done is calculated as:

$W_k = F_a\Delta x = 54N \times 11m = 590\,N-m\;or\;590\,joules$ Unless stated otherwise, it is understood that the object moved has a constant speed.

Example 2: Calculation of Work Using a Component of the Force

A force of 68 lb must be applied to the box in Figure 9.2 to move it across the surface in the diagram. The angle at which the force is applied is 42° with the horizontal and the distance the object moves is 12 ft. How much work is done? The analysis of the situation is the same as in Example 1 but we must determine the component of the force that is in the direction of the displacement in order to solve the problem. The best way to do this is by drawing the force to scale and measuring the angle with a protractor. Construct the projection of the force on a horizontal line to find the horizontal component of the applied force. I urge you to do that, and I give you my results for you to check. The horizontal component of the applied force is 51 lb. The solution is:

$$W_k = F_{AH}\Delta x = 51lb \times 12ft = 610\ ft - lb$$

Notice that the answer is given to two significant figures because the values given in the problem were given to two significant figures.

Example 3: Calculating Work Using Lifting Force

A woman lifts a 5.3 kg box from the floor of her garage and places it on a shelf 2.5 m above the floor. How much work did she do? This problem is a little different because the object is moved vertically. You know that if the box is caused to move upward at a constant speed near the surface of the earth, then the force applied must be the weight of the box. After reviewing Newton's laws, you may figure out how to calculate the weight of the box. These are ideas you must have in mind when you begin to construct a solution. The analysis of the problem into its parts will enable you to enjoy a quick solution to the problem.

$$W_k = F_A\Delta y = mg\Delta y = (5.3kg)(9.8m/s^2)(2.5m) = 130\frac{kg-m}{s^2} \times m = 130N - m = 130\ joules$$

These examples should serve you well as models for solving simple problems involving the calculation of work done. One other idea should be included to make your collection of tools for solving work problems more complete. I told you earlier that it is assumed that the object moves with a constant speed unless you are told otherwise. You have found that a lot of information is implied in the statement of a problem and that makes it interesting to solve. The way we can be sure that the object moves with a constant speed is to make sure there is no net force. That will be true if the force you apply is equal in magnitude but opposite in direction to any other force on the object. You did that in the solution to Example 3. The force applied upward to the box was equal to the weight that acts downward. No one will point these things out to you when you are confronted with a problem; that is why I am preparing you to handle the physics involved.

In Example 1, can you guess what the other force was that kept the object from accelerating? That's right, friction was acting in the direction opposite to the direction of the applied force. The force of friction was equal in magnitude, but opposite in direction, to the applied force. Now look at Example 2. The force of friction would not be in the direction of the applied force. I'll bet you already realized that it would act in a direction opposite to the horizontal component of the applied force. Forces acting on objects when work is done are not always constant. Graphical techniques or calculus are needed to handle those situations, but handling cases when forces are constant will prepare you to handle the other situations when they arise.

Any time you move something there will be an opposing force, the force of friction. In many cases, it is negligible and you do not have to account for it. In some problems, you will have to find the force of friction. Usually the force you will have to apply to get the object to move at a constant speed is equal in magnitude to the force of friction. There is a force of moving friction and a force of starting friction. You already know that it is harder to get something to start moving than it is to keep it moving. The moving friction is the force you should concern yourself with even though you are well aware that it may be a little less than the force you applied initially.

The force you apply at first has to get the object up to speed. Therefore, the object is accelerating. We are talking about the time after that initial acceleration when the object is moving at a constant velocity. The force of friction depends on a lot of things, such as the types of surfaces in contact, whether the surfaces are rough or smooth, whether the surfaces are bare and dry or lubricated in some way, and so on. Again you should be aware of how those things can change the force of friction, but for now you can concentrate on the bare and dry surfaces that cause an opposing force.

The two surfaces you will consider have a characteristic coefficient of friction. Usually the coefficient of friction is either given in a problem if it is needed, or enough information is given for you to calculate it. Sometimes a reference to a table will be given for you to find the coefficient of friction when you need it. Once you know the coefficient of friction for two surfaces, the force of friction is calculated as $\Im = \mu N$, and in words it means the force of friction is equal to the product of the coefficient of friction and the normal force. The normal force means the force perpendicular to the surfaces. If you look up the word in some dictionaries, you will find that one of the old definitions of the word *normal* is perpendicular. Kind of weird, I know. I bet you are getting nervous about now. I will relieve some of the stress just created by giving you another example.

Example 4: Calculating Work Using the Force of Friction

How much work is done to roll a metal safe, mass 121 kg, a distance of 20.5 m across a level floor? The coefficient of friction is 0.049. You want to move the safe at a constant speed and that means the force you apply must be equal in magnitude to the force of friction. In this case, the normal force is the weight of the safe and you know how to calculate that.

$W_k = F_A \Delta x = \Im \Delta x = \mu N \Delta x = \mu W \Delta x = \mu m g \Delta x$ The algebraic solution states that the work done is equal to the coefficient of friction times the weight of the safe times the distance moved.

$W_k = (0.049)(121 kg)(9.8 m/s^2)(20.5 m) = 1200\, joules$ Are you ready to go for it all by yourself? Go!

Physics Phun

1. A man dangling by his hands on the monkey bar flexes his biceps and raises his body until his head is well above the monkey bar. His mass is 5.22 slugs and his body is raised 61.1 cm. How much work did he do? Remember to express given information in the same system of measurement. In this case it is probably easier to change centimeters to feet and calculate the work done in ft-lb. Hint: 2.54 cm = 1 in, g=32 ft/s².

2. The coefficient of friction between a packing crate and the floor is 0.234. If the packing crate weighs 372 lb, how much work must be done in moving it 21.5 ft across a level floor at a constant speed?

3. A woman pulls a sled across a level field of snow by applying a force of 68 N to a rope attached to the sled. The rope makes an angle of 38° with the snow. How much work does she do pulling the sled across the field a distance of 52 m?

4. A 1.2 kg ball is swung in a circular path at a constant speed. The frequency of the ball is 2.5 cycles/s and the radius of the path is .98 m. How much centripetal force acts on the ball? How much work is done? Explain your answer. Hint: Think about how much displacement occurs in the direction of the force. Also think if there is any component of the force acting in the direction of the motion.

5. A man pushes on the handle of a lawn mower with a force of 55 lb, and the handle makes an angle of 45° with the ground. How much work is done when he pushes the mower along a level stretch of 21 ft? Hint: The force is enough to keep the mower moving at a constant speed. The force is applied downward along the handle of the mower.

Making Work Easy

The title of this section sounds like an oxymoron but you have probably discovered a few ways to make work easier for yourself. You probably drive or ride a bicycle to work and use the stairs or a ramp instead of jumping straight up to get to an upper level of a building. These are some of the ways humankind has devised to do work easier.

There are six simple machines that enable us to make our work easier. Sometimes we combine them to make complex machines using two or more of them. I see the simple machines as part of two groups. The first group is made up of the lever, the wheel and axle, and the pulley. The second group includes the inclined plane, the screw, and the wedge. I will discuss one member of each family because all other members in the family have similar characteristics. But first I'll discuss general characteristics common to both groups.

In an ideal world, the simple machines would work with no loss in the work you put in compared to the work you get out. Any time we use a simple machine, we know that friction opposes any motion that may be involved. Another way of stating the ideal situation is to compare the work we get out of a machine with the work we put into it. As you already know, a ratio of two quantities is a good way to compare them. Ideally $\frac{W_{k\,out}}{W_{k\,in}} = 1$, but in our world, we know $\frac{W_{k\,out}}{W_{k\,in}} < 1$. That means the work you get out in the real world is less than the work you put in. Hey, wait a minute. Did I just say that you do more work using the machine? Be sure that I explain why you would want to do that before this chapter is done. You can express the inequality by introducing *efficiency*.

The efficiency is expressed as: efficiency $= \frac{W_{k\,out}}{W_{k\,in}} \times 100\%$. You see that efficiency is expressed in such a way that you know the efficiency of any machine will never be 100%. The electrical transformer is about 98% efficient, and that is about the upper limit of efficiency.

Plain English

The **efficiency** of a machine is a fraction greater than zero and less than one that expresses what part of input work the output work amounts to. The efficiency is usually stated as a percent obtained by multiplying the fraction by 100%.

Plain English

Actual mechanical advantage is the actual gain in force that a user realizes from a machine because it includes the effects of friction. **Ideal mechanical advantage** is the theoretical gain a user expects from a machine. It is always more than the AMA because it does not include the effects of friction.

There is some more funky language you should be familiar with to do a complete job of discussing simple machines. Since a machine is never perfect, something must be sacrificed when a machine is used. Distance or speed is gained by exerting considerable force. Riding a bicycle is a good example of sacrificing force to gain speed. Heavy weights may be lifted but only at the expense of distance and/or speed. Your auto mechanic can lift the motor from a car with a pulley system by applying a small effort force, but large amounts of chain links must pass through the pulley to lift the motor a small distance. Two terms are used to avoid explaining the limitations of a machine when it is employed. One term is *actual mechanical advantage*, and is represented by the letters AMA. The AMA is the real gain a user realizes when using a machine. The AMA is calculated as a ratio of forces as follows: $AMA = \frac{F_R}{F_e}$. Similarly, the *ideal mechanical advantage*, IMA, is the expected advantage if you did not have any friction.

The effort force, F_e, is the force applied to the machine. The resistance force, F_R, is the force the machine applies to an object.

The distance, S_e, is the distance through which the effort force moves. The distance, S_R, is the distance through which the resistance force moves.

The ideal mechanical advantage is calculated as: $IMA = \dfrac{S_e}{S_R}$. Either or both of these ideas may be used to solve problems involving machines. Both can be used to discuss efficiency because they have their origin in the calculation of work. You can write the ratio of output work and input work in terms of forces and distances as follows:

$\dfrac{W_{k\ out}}{W_{k\ in}} = \dfrac{F_R S_R}{F_e S_e} = \dfrac{\frac{F_R}{F_e}}{\frac{S_e}{S_R}} = \dfrac{AMA}{IMA}$. You can see what I mean about these new terms being very closely associated with our earlier discussion. In fact, efficiency $= \dfrac{AMA}{IMA} \times 100\%$. With these general properties of all simple machines in mind, we investigate one member from each of the groups of simple machines identified earlier.

The Inclined Plane

The inclined plane is a simple machine that is in wide use today and has many applications. The inclined plane makes it easier for you to lift a heavy object from one level to another with force less than the object's actual weight because it is some form of a ramp. It may be in one of many forms such as stairs or streets. Refer to Figure 9.3 as we discuss the details of an inclined plane. The inclined plane has length l and height h.

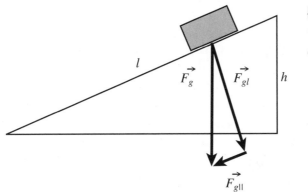

Figure 9.3

The inclined plane is one example of a simple machine.

The inclined plane can be used to raise the object vertically a distance h by applying a force parallel to the plane for a distance l rather than lifting the object straight up.

The magnitude of the weight of the object is F_g, which has been resolved into components parallel to l and perpendicular to l, $F_{g\parallel}$ *and* $F_{g\perp}$ respectively.

The weight of the object is F_R. If you wish to pull the object up the plane at a constant speed, then you must apply F_e parallel to l and up the plane opposite $F_{g\parallel}$. The magnitude of F_e is the sum of $F_{g\parallel}$ and the force of friction. That is, $F_e = F_{g\parallel} + \Im =$ $F_{g\parallel} + \mu\, F_{g\perp}$. The length of the inclined plane is S_e and the height is S_R. You will determine the values of $F_{g\perp}$ and $F_{g\parallel}$ when you construct the force triangle and resolve the weight into two mutually perpendicular components. You will then be able to calculate

$$\text{AMA} = \frac{F_R}{F_e} = \frac{F_g}{F_{g\parallel} + \mu F_{g\perp}} \text{ and IMA} = \frac{S_e}{S_R} = \frac{l}{h}.$$ You can solve any inclined plane problem with the information outlined here.

Problem:

An inclined plane is 8.00 ft long and 2.00 ft high. How much force must be applied parallel to the plane to pull an object weighing 2020 lb up the inclined plane at a constant speed if the coefficient of friction is 0.049? What is the efficiency of this machine? A construction of the inclined plane to scale and resolving the weight into two mutually perpendicular components yields the following information: $F_{g\parallel}$ = 506 lb, $F_{g\perp}$ = 1960 lb, $S_e = 8.00$, $S_R = 2.00\, ft$, $\Im = \mu N = (.049)(1960 lb) = 96 lb$. The effort force, $F_e = F_{g\parallel} + \Im =$ 506 lb + 96 lb = 602 lb.

Solution:

The efficiency of this machine can be calculated as:

$$\frac{AMA}{IMA} \times 100\% = \frac{\dfrac{2020 lb}{602 lb}}{\dfrac{8.00\, ft}{2.00\, ft}} \times 100\% = \frac{3.36}{4.00} \times 100\% = .84 \times 100\% = 84\%.$$

This machine is 84% efficient, and that means you get back 84% of the work you put into the machine. You still need three or four good men to pull the load up the inclined plane, but many more good men to lift the load two feet straight up. For those of us who are weak of body but strong of mind, we prefer the use of an inclined plane to the snatch and lift technique. Remember that I told you there would be a reason for doing more work using the machine. When the force is too big for you and your friends alone to pull or push an object straight up, use an inclined plane. You may do more work, but the job of lifting will get done.

The Lever

The lever is a member of the other group of simple machines that deserves our attention. An understanding of the lever reveals the basic principles of all other members of that group. The lever, like the inclined plane, has associated with it the ideas represented by the symbols F_R, F_e, S_R, S_e, *IMA*, and *AMA*.

The effort force, the force applied by the user of the machine, is the same for the lever as it was for the inclined plane. The resisting force is the same as well, the force the machine works on is often the weight that is being lifted. In order to understand S_e and S_R, which play the same role for the lever as the inclined plane, you need to realize that S_e and S_R are arcs along circles. Furthermore, $S_e \propto l_e$ and $S_R \propto l_R$ are labeled for you in Figure 9.4.

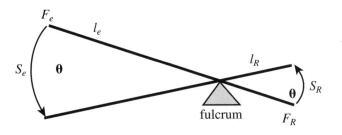

Figure 9.4

A lever as a simple machine showing necessary parts.

The output part is on the right side of the diagram and the input on the left. A lever pivots about a point called the fulcrum. There are three different types of levers. The one used in the diagram lifts a large force, or weight, with a smaller effort force. You could do it in reverse. That would be the case if you wanted to gain distance instead of force. We will save that for another problem later. In the diagram, the effort force is applied to the left end that moves through a circular arc labeled S_e that represents the effort distance. The resistance force is applied to the end of the lever on the right in the diagram and that end moves along the circular arc labeled S_R. The distance moved by the effort force is calculated in terms of l_e. Similarly, the resistance force moves a distance that is expressed in terms of l_R.

$$IMA = \frac{S_e}{S_R} = \frac{l_e}{l_R}$$

The lever shown in the diagram is used to multiply force. It would have served to help lift the SUV from Chapter 5 out of the mud hole. This same lever can be used to multiply distance if the effort force is applied at the other end. An example of a lever used to multiply distance is the use of scissors. Some levers have the fulcrum at one end and the effort force is applied at the other end, such as an oar for a boat where the blade is the fulcrum. The same arrangement can be used where the resistance force is on the end opposite the fulcrum and the effort force is applied between the fulcrum and the resistance force.

The lever, like any other simple machine, will multiply force or multiply distance. Even though you know that a machine cannot do both in our world because of friction, you can often assume friction is negligible in the solution of many problems unless enough information is given to prove otherwise. Friction can be good for us, as you know. It would be impossible to roll a wheel or to take a step walking, to mention a couple of good uses of friction.

Problem:

A board 8.3 m long is used as a lever with the fulcrum between the resistance force and the effort force. The fulcrum is placed 3.3 m from the resistance force. (a) Calculate the IMA of the simple machine. (b) What is the minimum effort force that should be applied to lift the weight of a 55 kg mass?

Solution:

You know that the IMA is the ratio of resistance and effort arms. That is,

(a) $IMA = \dfrac{l_e}{l_R} = \dfrac{5.0m}{3.3m} = 1.5$. Since we do not have enough information to determine the efficiency of the machine, we can figure out the force needed if we assume that it is 100%. That would be the minimum force needed. If there were friction, it would just mean that the actual force would have to be more than that. That means,

(b) $AMA = \dfrac{F_R}{F_e}$ and $F_e = \dfrac{F_R}{AMA} = \dfrac{(55kg)\left(9.8\dfrac{m}{s^2}\right)}{1.5} = 360N$. What if you had multiplied by 1.5 instead of dividing? You would have given an answer that would say that it takes more force to lift the object than it weighs. This can't be right if the fulcrum is nearer to the object than the force being applied by the lifter. Thinking about the problem to see if your answer makes sense can really help you when you work problems.

Physics Phun

1. A crate weighs 555 N and is pulled upward along a 31 m inclined plane at a constant speed to raise it 2.3 m above the starting point. The force required to pull it along the plane is 55.5 N. Calculate (a) the work output, (b) the work input, (c) the efficiency of the machine.

2. A holiday nut is placed 2.54 cm from the hinge of a nutcracker. If you exert a force of 51 N at a point 15.2 cm from the hinge, what resistance force does the nut exert?

Watt Is Power?

Is that a play on words? You may not know watt is true. The speed demon would probably say horsepower is power, and that is correct also. At least they are both units of power, as is shown in this section. Be aware of some things that are said about power that we should be able to respond to with confidence. For instance: Does the power company sell power? Is the more powerful car the one that can lay down the most rubber when it leaves the parking lot? These ideas will be addressed because they fall within the scientific notion of power. Just as work has a different scientific meaning than the general notion of work, so does power. The first thing we should note is that power is a scalar quantity. There is no direction associated with it.

Power is the work done per unit of time. Some textbooks define power as the rate of doing work. Power depends upon force applied in the direction of displacement, the displacement, and time. The definition of power in symbols is $P = \dfrac{W_k}{t}$. As we always do, consider the units of power in all three systems of measurement.

In the CGS system, the units of power from the definition are $P = \dfrac{ergs}{s}$. There is no nice, neat unit for power in that system except ergs/s. The MKS system units of power are a little more meaningful: $P = \dfrac{joule}{s} = watt$. The units joule/s are always units of power in the MKS system and $1\dfrac{joule}{s} = 1 watt$. You are probably more familiar with the kilowatt or 1000 watts. As usual, the FPS system is weird. The units of power by the definition are $P = \dfrac{ft - lb}{s}$ in the FPS system. You are probably more familiar with another unit of power in that system, the horsepower or hp. The horsepower is a fairly large unit because $1 hp = \dfrac{550\, ft - lb}{s}$. So if you have 300 horses under the hood of your car, you have a car that will do 165,000 ft-lb of work every second! That is real power.

You noticed the 165,000 ft-lb of work every second. That is what your power company sells to you in the form of electricity. The thing you pay the power company for is work, not power. You pay the electric company based upon your kilowatt-hours. These are units of work and not power. Look at the units. You are multiplying a unit for power times a unit for time. Now look at the equation that defines power. If you multiply both sides by time, you get Pt = W. So kilowatt-hours are units of work. Suppose you calculate the work required for a 100 watt bulb to provide light in your room. You can say that every second the amount of work is 100 watts of work. In an hour, the bulb requires $(100 watts)(3600 s) = 3.6 \times 10^5 watt - s = 1 \times 10^2 watt - hr = 1 \times 10^{-1} kilowatt - hr$ or $0.1 kw - hr$.

You now have the information you need to compare the power of two cars. You have decided already that the car that lays down the most rubber when starting off is not necessarily the most powerful. If you have two identical cars climb the same hill and time them, the car that gets to the top in the shortest time is the more powerful. They both do the same amount of work on the inclined plane but by the definition of power, the smaller the time the greater the power.

Problem:

A man's mass is 82.0 kg. If he walks up a flight of stairs 3.0 m high in 9.0 s, what power has he used?

Solution:

M = 82.0 kg, t = 9.0 s, h = 3.0 m, P=?

$$P = \frac{W_k}{t} = \frac{(82.0\,kg)(9.80m/s^2)(3.0\,m)}{9.0\,s} = 270\,watts.$$

The Least You Need to Know

- ◆ Use the scientific definition of work.
- ◆ Identify the units of work in each system of measurement.
- ◆ Relate work to simple machines.
- ◆ Describe simple machines in terms of IMA, AMA, and efficiency.
- ◆ Use work to define power and identify the units of power.

Chapter 10

Kinetic Energy

In This Chapter

- ◆ Describing kinetic energy
- ◆ How does work enter the scheme?
- ◆ Raindrops and kinetic energy
- ◆ Following kinetic energy
- ◆ Kinetic energy and air pressure

My physics students told me that the physics course itself is not as difficult as they anticipated after they got into it. They did feel that words like kinetic energy are scary the first time they are confronted by them. I made an effort to help them adjust to new concepts so that they felt comfortable discussing them with each other and with me. I make the same effort to help you in this chapter because I know that it is no fun to run into something brand new when no one is around to help you deal with it.

What Has Kinetic Energy?

This is the first time that I have mentioned energy and that can be unnerving reading about it in a book about physics. Even though you read about it in the newspaper or see reports about it on television daily, chances are you think of it as just another topic of conversation by the media. When you are about to venture into a discussion of the same topic in physics, that is something else.

The topic of our discussion is involved in something that you have watched on news reports, in movies, and maybe even firsthand. Here are some examples:

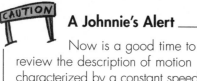

A Johnnie's Alert _____

Now is a good time to review the description of motion characterized by a constant speed as well as motion characterized by constant acceleration. The discussion of kinetic energy will require knowledge of the description of motion of an object.

- A hurricane hitting the coast crashing giant waves into barriers; trees bending in the wind and roofs of homes flying away like leaves.

- A towering punt with an unbelievable hang time; a receiver running under the nearly one pound football, catching it, and running.

- A medicine ball passed from person to person.

- A swerving unguided missile in the form of a giant, free-rolling truck wheel.

Plain English _____

Kinetic energy is the energy an object has when it is in motion.

Each of these examples has what scientists call _kinetic energy_. Kinetic energy is the energy an object has due to its motion. You see that some new ideas may be scary at first, but when you realize that you have lived with the phenomena most of your life it is really quite comforting. It is time to roll up your sleeves and dig in.

Work and Kinetic Energy

It is probably no surprise for you to find that work and kinetic energy are related in some way. I will try to stay in familiar territory by using ideas introduced so far in this book. I review some concepts needed to accomplish the task. In this context, I use v for initial velocity and v' for the final velocity. Since we are discussing motion and force in a straight line in the same direction, we won't bother with vector notation.

The motion is uniformly accelerated motion so we have three ideas to guide us in our analysis, along with Newton's second law to help us with the force due to gravity causing the wheel to roll merrily along down the inclined plane. We solve algebraically in terms of the quantities involved.

$$\frac{v + v'}{2} = \bar{v}, \Delta x = \bar{v}t, a = \frac{v' - v}{t}$$ Information we know for describing uniformly accelerated motion.

$$\Delta x = \bar{v}t$$

$$\Delta x = \bar{v}\left(\frac{v' - v}{a}\right)$$

$$\Delta x = \left(\frac{v + v'}{2}\right)\left(\frac{v' - v}{a}\right)$$

$$\Delta x = \frac{v'^2 - v^2}{2a}$$

This is a description of the motion of the orphaned wheel in terms of things we could observe: initial velocity, final velocity at its destination, acceleration, and a distance traveled before arriving at a safe destination. Now the task is to find a way to include the force and look for a relationship with work: $a\Delta x = \frac{v'^2 - v^2}{2}$.

Now what? If we only had mass ... the wheel had mass, so multiply both sides of the last equation by m: $ma\Delta x = \frac{mv'^2 - mv^2}{2}$

That will do it! Remember that mass times acceleration is equal to the net force. Notice that the left member of the equation, therefore, is work—that is, $F\Delta x = \frac{mv'^2}{2} - \frac{mv^2}{2}$. The quantity $\frac{1}{2}mv^2$ is the kinetic energy. The last equation in our derivation states that the net work done is equal to the change in kinetic energy. In symbols, $W_k = \Delta KE$, where $KE = \frac{1}{2}mv^2$.

This relationship gives us a lot of information about kinetic energy. Kinetic energy is a scalar quantity. The units of kinetic energy are the same units as force times distance in all systems of measurement. Since the net work done is equivalent to the change in kinetic energy, then the change in kinetic energy is equivalent to the net work done by a crashing runaway wheel! Given enough information, you can calculate the amount of work done on that wheel to get it to come to rest. That kind of work can do a lot of damage. It is time to look at some applications of this new idea.

CAUTION

A Johnnie's Alert

Many problems state given information explicitly while other information is implied. When an object is dropped from a point above the surface of the earth, the initial velocity is zero, and the acceleration is the acceleration due to gravity.

Example 1: Motion on an Inclined Plane

A crate, mass 2.00 kg, slides from rest down a frictionless inclined plane. The inclined plane is 4.00 m long and makes an angle of 15° with the horizontal.

(a) How large is the force that accelerates the crate down the plane?

(b) How fast is the crate traveling when it reaches the bottom of the plane?

(c) How much time does it take the crate to slide down the plane?

![CAUTION] **A Johnnie's Alert**

Resolving a given vector into components is often implied in the statement of a problem. You will decide whether or not finding the components is necessary as well as which components are needed. A large percentage of cases will involve components of a vector that are mutually perpendicular.

It is a good idea to sketch a diagram and label it in terms of the information given and the information you are asked to find. I do that and include it as Figure 10.1. Refer to that diagram as you consider the solution. Part of the solution is constructing the diagram correctly and labeling each part possible. I drew a horizontal line to represent the base of the inclined plane and constructed a 15° angle at one end. I then extended the one side of the angle to represent 4.00 *m*; use any scale you choose.

Figure 10.1

The inclined plane for Example 1 includes important labels.

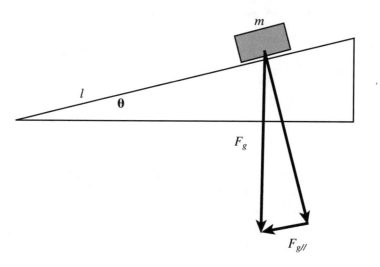

I then constructed the weight of the crate, F_g, and resolved it into components parallel and perpendicular to the plane (the side *l* is referred to as the plane in the context of problems dealing with inclined planes). I labeled each part and am now ready to construct the solution.

$m = 2.00kg$ $v = 0$

$l = 4.00m$ $\Im = 0$

$\theta = 15.0°$

(a) $F_{g\parallel} = ?$

(b) $v' = ?$

(c) $t = ?$

(a) $F_g = mg = 2.00kg \times 9.80\dfrac{m}{s^2} = 19.6N$ You can use this value of the weight to construct a vector to represent 5.07 N. When you construct the force triangle as I have in Figure 10.1, the geometry of the triangle will dictate the length of $F_{g\parallel}$ so that when you measure it using the same scale as you used to construct F_g, you will find that $F_{g\parallel} = 5.07N$.

(b) When you analyze the next part, you know that the motion down the plane is uniformly accelerated motion, and you can solve for acceleration in terms of the mass and the net force, the component of the force of gravity parallel to the plane. However, to find the final velocity we need the time to travel the length of the plane. There must be another way. Fortunately there is, and we have just developed the relationship between kinetic energy and work. Try this idea:

$$W_k = \Delta KE$$

$$F\Delta x = \frac{1}{2}mv'^2, \text{ since } v = 0,$$

$$F_{g\parallel}\Delta x = \frac{1}{2}mv'^2$$

$$v' = \sqrt{\frac{2F_{g\parallel}\Delta x}{m}}$$

$$v' = \sqrt{\frac{2(5.07N)(4.00m)}{2.00kg}} = \sqrt{20.3\frac{m^2}{s^2}} = 4.50m/s$$

(c) Once you have the final speed, you can solve for t.

$$a_{\parallel} = \frac{v' - v}{t}$$

$$\frac{F_{G\parallel}}{m} = \frac{v'}{t} \quad \text{Since } v = 0 \text{ and } \Im = 0.$$

$$t = \frac{mv'}{F_{g\parallel}}$$

$$t = \frac{(2.00kg)(4.50)m/s}{5.07N} = 1.78s$$

Notice that this problem requires the use of the following ideas: (a) uniformly accelerated motion, (b) Newton's second law of motion, (c) work, (d) kinetic energy, and (e) the relationship between work and kinetic energy.

A Johnie's Alert _____

Check your understanding of the calculation of work as well as the units of work in each system. Having this information readily at hand will often make the solution of a problem obvious. Read Example 2 carefully before planning a strategy for a solution. Using this alert may give you a big surprise as you consider a solution to that problem!

Example 2: Kinetic Energy of an Object Moving in a Circular Path

An object of mass 0.500 kg is moving in a circular path of radius 1.00 m at a constant speed with a frequency of $0.500\ s^{-1}$.

A Johnnie's Alert _____

When you read a physics problem, make note of the units of measurement of each item of given information. If all of the information is not in one system of measurement, choose a system and use unit analysis to convert all measurements to your choice of CGS, MKS, or FPS.

(a) What is the centripetal force?

(b) How much work is done?

(c) What is the kinetic energy of the object?

This problem requires some thought as you work through it. The first part is fairly straightforward because it just requires an expression for centripetal force. Since we are given the frequency, the mass, and the radius, the centripetal acceleration can be calculated then multiplied by mass; we may use

$$F_c = \frac{m4\pi^2 R}{T^2} = m4\pi^2 Rf^2 = (0.500kg)(4\pi^2)(1.00m)(0.500s^{-1})^2 = 4.93N$$ as a solution

to (a). Do you remember this problem from a previous chapter? There are no tricks or magic, but you do have to think critically. Since there is no displacement in the direction of the centripetal force (radially inward toward the center), no work is done by the applied force! To solve (c), we calculate the KE as

$$KE = \frac{1}{2}mv^2 = \frac{1}{2}(0.500kg)\left[(2\pi)(1.00m)(0.500s^{-1})\right]^2 = 2.47\frac{kg-m^2}{s^2} = 2.47N - m = 2.47\,joules.$$

Notice that I have observed the rules for scientific notation and significant figures in both examples. I know you can't wait to work on your own for a while so here goes!

Physics Phun

1. What is the kinetic energy of a baseball, weight 5.1 oz, thrown with a speed of 93 mi/hr?

2. The speed demon's car, weighing 2208 lb, accelerated on a level road from 40.0 mi/hr to 60.0 mi/hr in 11.5 s. (a) Calculate the change in kinetic energy. (b) Calculate the force that produced the acceleration. (c) How much power, in horse-power, was developed?

3. An electron is moving at a constant speed of 1.00×10^9 *cm/s*. Calculate its kinetic energy if the mass of the electron is 9.11×10^{-28} *g*.

Kinetic Energy of Falling Objects

Most of you have heard of hailstones falling from the sky making terrible dents in cars; many of you perhaps have seen this firsthand. You know that hailstones have a tremendous amount of kinetic energy. We will discuss the kinetic energy of falling objects briefly and then take the idea up again in the next chapter.

You know that any time something is in motion it has kinetic energy. If we drop an object from some point above the surface of the earth, you know that it is accelerated by the force of gravity and continues to gain speed until it reaches terminal velocity, hits the ground, or dents a car. The fact that it is dropped means that the initial speed is zero. Therefore, the kinetic energy of any falling object that was dropped is given by

$$KE = \frac{1}{2}mv'^2.$$

Suppose that one of the hailstones had a mass of 4.00×10^{-4} slugs and is dropped from a cloud 5.00×10^2 ft above the earth. Use your knowledge of falling objects to calculate the speed of the object when it hits the earth. I work this with you here:

A Johnnie's Alert

Motion along a straight line in one direction can be described by using the terms velocity and speed interchangeably. In addition, make sure you can express work in fundamental units as well as joules, ergs, and ft-lb.

$-h = 5.00 \times 10^2$ ft   the distance the object falls

$a = -g$ the acceleration of gravity $v = 0$ Because the object is dropped.

$$-h = \bar{v}t = \left(\frac{v'+v}{2}\right)\left(\frac{v'-v}{-g}\right)$$

$$-h = \frac{v'^2 - v^2}{-2g}$$

$$2gh = v'^2$$

$$v' = \pm\sqrt{2gh}$$

$$v' = \pm\sqrt{2(32.0\,ft\,/\,s^2)(5.00\times10^2\,ft)}$$

$v' = -179\,ft\,/\,s$ Remember that the minus sign means it has a downward velocity because it is falling.

Calculate the kinetic energy of the falling object and you get:

$$KE = \frac{1}{2}(4.00\times10^{-4}\,slugs)(-179\,ft\,/\,s)^2 = 6.41\,ft - lb.$$

Remember the stone you dropped from the bridge and saw a splash 3.00 s after you dropped it? You found the height of the bridge in that problem. Suppose we do the same experiment but this time we calculate the kinetic energy just as the stone hits the water if it has a mass of 5.00 g.

$$-h = \frac{v'^2 - v^2}{-2g}$$ From the last solution.

$$-h = vt - \frac{1}{2}gt^2$$ Solving for –h in terms of t, v, and –g.

$$-h = -\frac{1}{2}gt^2$$ Since v = 0.

$$-\frac{1}{2}gt^2 = \frac{v'^2 - v^2}{-2g}$$ Substituting the last equation into the first.

$$g^2t^2 = v'^2$$ Since v = 0.

$$KE = \frac{1}{2}mg^2t^2$$

$$KE = \frac{1}{2}(5.00g)(980cm\,/\,s^2)(3.00s)^2 = 2.16\times10^7\,\frac{g-cm^2}{s^2} = 2.16\times10^7\,ergs$$

Physics Phun

1. Suppose a bowling ball of mass 6.82 kg is dropped from a point 103 m above the earth. (a) Calculate the speed of the bowling ball just before it touches the earth. Did you list all of the information you are given and what you are to find? Did you solve the problem algebraically first? (b) Calculate the kinetic energy of the ball just before it touches the earth. (c) What would you expect to happen to the kinetic energy after the ball touches the earth?

2. A 2000 lb car traveling at 60 mi/hr came to a safe sudden stop. (a) What was the change in kinetic energy of the car? (b) How much work was done on the car? (c) What do you think caused the change in kinetic energy?

Where Does Kinetic Energy Go?

Did it occur to you that if the kinetic energy an object has is a result of its motion, where does the energy go when the object is stopped? A more complete answer will be clear after the next chapter but we can tackle part of that question now.

Remember the hailstones? What did they do when they hit the car? They dented the car and bounced off. However, the hailstones were not traveling as fast as they were when they first fell. What do you suppose the bowling ball in the last set of Physics Phun problems did after it hit the ground? It probably made a dent in the earth and maybe bounced a tiny bit. Something happened to the kinetic energy, because there is just a muddy bowling ball lying there after a small bounce or after it became stuck. What about the speed demon's car in the last problem set? What if instead of stopping safely, it ran into a post? There would have been some vicious car reshaping that would have taken place.

The thing that happened in each of those situations is what you expected: work was done on the object to stop it. That is just part of the explanation because if you felt the dent made by the hailstone or bowling ball, it would probably feel warm. If you cannot imagine that, consider pounding a nail into wood with a hammer. If you feel the nail, there is definitely a temperature increase generated from the loss of kinetic energy of the hammer, and work was done to drive the nail into the wood. The increase in temperature indicates that there is thermal energy in the nail as a result of the transfer of kinetic energy.

So part of the story is that some of the kinetic energy goes into the distortion of the wood and some into heat. Some of the loss of kinetic energy seems pretty tragic, and it is in some cases. You do not like to think of crashing cars even though that happens every day. But there are situations where this can be most helpful. You probably have seen, and heard, the operation of a pile driver. A huge engine lifts the large hammer and drops it on the pile and work is done driving the wooden post into the ground. If you hear it pounding, you know that some of the kinetic energy is changed to sound. That is the topic of Chapter 16, so you can return to the idea later.

The Kinetic Energy of a Projectile

It is interesting to consider kinetic energy in the context of a problem discussed earlier. Please refer to Figure 7.2 where we considered the path of a projectile (see Chapter 7). Suppose that $\vec{v_0} = 30.0 m/s$ at an angle of 40.0° with the horizontal and its mass is 114 g. Calculate the kinetic energy of the tater initially, at maximum height, and just before it touches the ground. Here again, we will be able to detect any losses of kinetic energy, if there are any, but we must wait until later chapters to get a more satisfying answer to the problem.

Remember that in physics if you are told the whole truth about one phenomenon you probably will be stuck at that place forever. Enough of the truth to complete part of the job now is great and anticipating the rest of the story gives you something to look forward to.

I begin a solution for the problem of the airborne potato by realizing that everything necessary to construct a solution is not stated explicitly. I need to resolve the initial velocity into vertical and horizontal components. I do that by constructing a line tangent to the path initially at an angle of 40° with the horizontal. You know that the trajectory changes with different angles associated with the initial velocity. The horizontal component is 23.0 *m/s* and the vertical component is 19.3 *m/s*.

A Johnnie's Alert

Pick a solution to a physics problem that you understand and one that ties a lot of ideas together for you to use as a model for your study time. The path of a projectile is such a problem.

You know that the projectile involves two types of motion simultaneously. That is why we know that the acceleration of the projectile at maximum height is 9.80 *m/s*² and the vertical velocity is zero. However, it still has its horizontal velocity. Use that velocity to find the kinetic energy.

The kinetic energy at maximum height is then $KE = \frac{1}{2}(0.114kg)(23.0m/s)^2 = 30.2\,joules$. The kinetic energy initially is $KE = \frac{1}{2}(0.114kg)(30.0m/s)^2 = 51.3\,joules$. That means there was a loss of kinetic energy from the initial launch to the point where the tater reaches maximum height. Where did it go? You probably already know but we will play the game anyway.

After completing the next chapter, you will be able to do a better job of accounting for the kinetic energy of a projectile in flight. Just before the projectile touches the ground, the instantaneous velocity is –30.0 m/s because it is going down. The kinetic energy at that time is $KE = \frac{1}{2}(0.114kg)(-30.0m/s)^2 = 51.3\,joules$. The kinetic energy has returned! If you don't have an explanation for what is happening here, maybe this has piqued your curiosity and you have something to look forward to.

Bernoulli's Principle

This is another topic that will be made clear to you later but we need to begin exploring some ideas like this sometime. There is no time like the present to explore a new situation that involves kinetic energy. Appealing to your general knowledge before we begin, you are aware of air pressure. I plan to spend a lot of time with you in Chapter 14 discussing fluid pressure in detail. All you need to know now is a few basic things about air pressure. For instance, when you drink liquid through a straw, you lower the air pressure inside the straw, and the atmospheric pressure on the liquid outside the straw causes the liquid to rise in the tube to your mouth. Don't you think that sounds unappetizing? Anyway, I think that you know about pressure as it is related to sipping a milkshake through a straw.

Another case you are probably familiar with is the operation of a vacuum cleaner. The vacuum cleaner removes air from above the rug, and the atmospheric pressure under the rug pushes the dirt and dust up into the vacuum cleaner. These are oversimplifications, but I just wanted to review some general information that you can have at your fingertips that can be used to build some new ideas.

A Johnnie's Alert

The figures in this book are drawings I made with simple geometric drawing instruments or on the computer. Making a drawing and labeling it correctly is an important part of many physics problems. Practice making sketches and diagrams for yourself to help you construct solutions to problems.

You know that kinetic energy is directly proportional to the square of the velocity. That means that, all other things being equal, the kinetic energy quadruples as the velocity doubles. You can see that the kinetic energy increases rapidly with increases in velocity.

Bernoulli's principle deals with fluids that have a nice smooth flow for equal volumes. The nice smooth flow is called a streamline flow. Cars and other vehicles are built with the streamline shape to allow fluids to move around them with the streamline flow. Refer to Figure 10.2 where you will see the cross section of an airplane wing built so that air can move around it in a streamline flow. I have marked the path of some air particles that pass under the bottom of the wing in nearly a straight line. There are a couple of paths of particles that flow over the top. I exaggerated those paths to emphasize to you that both paths are traveled by air particles in the same amount of time. That means that the particles on top of the wing are traveling faster than those along the bottom of the wing.

Figure 10.2

The diagram of air particles moving around the cross section of an airplane wing illustrates Bernoulli's principle.

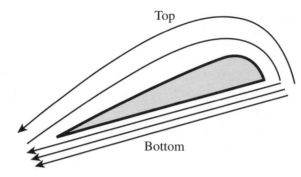

Top

Bottom

Bernoulli's principle states: $\frac{KE}{V} + p =$ constant. Stated in words, the kinetic energy per unit volume plus the pressure is a constant. That means that the air particles going over the top of the wing in Figure 10.2 are traveling much faster than those under the bottom and the kinetic energy per unit of volume of air must be larger than on the bottom. If $\frac{KE}{V} + p =$ constant, and $\frac{KE}{V}$ is larger, then p must be smaller if the sum of the two are to always be the same constant. Therefore, the pressure on top of the airplane wing must be less than on the bottom. The result is that the atmospheric pressure will tend to push upward on the bottom of the plane toward the region of lower pressure. It is much like the atmospheric pressure pushing liquid into the straw where you have reduced the pressure inside the straw by removing the air.

Bernoulli's principle applies to liquids as well as gases like the air. The most dramatic effects that we observe are the effects like those just discussed. If you can do anything to change the pressure of the streamline flow of a fluid by changing the speed of some of its particles, you will observe the effects just described. You will revisit this idea when we discuss pressure in more detail in the section "Does a Baseball Curve?" in Chapter 14.

You will be able to see that the difference in pressure above and below the wing of an airplane provides the lift necessary to keep the plane in the air along with the thrust of the motor. But that comes later. The important thing for you to notice here is the necessity of knowledge of kinetic energy before a concept like Bernoulli's principle is discussed with any understanding at all. No matter whether you are considering an airplane wing moving through an air mass, hailstones falling from the sky, or medicine balls and footballs being hurled at you by well-meaning friends, you are observing objects that have kinetic energy.

The Least You Need to Know

- Describe the kinetic energy of an object.
- Calculate the kinetic energy of an object.
- Relate the kinetic energy of an object to work.
- Detect the loss or gain of kinetic energy.
- Recognize units of measurement of kinetic energy.

11

Potential Energy

In This Chapter

- ◆ Sometimes it is potential energy
- ◆ Then it becomes work
- ◆ Anything that goes up has it
- ◆ Potential energy can be shared

You now know that there are several names for energy. The study of energy is an important part of your study of topics in physics. The current focus is on mechanical energy. Mechanical energy is often classified as kinetic energy or potential energy. In the last chapter you found that I described kinetic energy rather than defining it; I did that because defining kinetic energy is almost impossible for me since I cannot define energy.

As was noted before, we could say it is the ability to do work, but to me that is a cop out. So we will be satisfied for the time being to describe potential energy as well, then relate it to something that you understand.

When Does an Object Have Potential Energy?

Did you realize that no one knows what a point and a line are except for a "concept" in our minds? Geometry identifies these quantities as "undefined terms." Furthermore, all other "terms" in geometry are defined in terms of point and line. Along with axioms and postulates, the whole subject of geometry is built on the undefined terms point and line. Maybe terms like energy, mass, time, and space are to physics what point and line are to geometry. Strange, I agree with you, but please keep these ideas in mind. Maybe we are safe identifying properties and then talking about physical objects as having those properties without our knowing how to define the properties.

You have been exposed to more than enough of that for now. We described kinetic energy as the energy a body has due to its motion. In the same way, I describe *potential energy* as the energy an object has as a result of its position in a force field. You know about the gravitational force field, so we can begin with that. Remember that a force field is a region of space where a force acts at a distance on an object placed anywhere in that region. Before we start, I caution you about the relative nature of potential energy.

Plain English

Potential energy is the energy an object has because of its position in a force field.

An object has potential energy with respect to some level of reference. Usually the lowest level involved in a given situation is chosen as the zero level of potential energy. In this section, we will choose ground level, or the level with zero potential energy, as our reference point or level. That means that anything having mass and located at a position above ground level has positive potential energy. Anything at ground level, zero potential energy, and anything below ground level, negative potential energy relative to ground level. You will find the very same idea when working with electrical circuits; the ground has zero potential. Circuits also have positive and negative potential energy. Our work here with potential energy prepares you to deal with the same notion in other contexts.

Work and Potential Energy

You now have a description of potential energy and I can relate potential energy to an idea familiar to you. I will use the diagram in Figure 11.1 to establish that relationship among the quantities work, kinetic energy, and potential energy. Please notice that I have marked the zero level (0) for the level where potential energy is zero on the diagram.

Suppose the object of mass m initially at the foot of the cliff is lifted straight up to the top of the cliff by a force equal in magnitude to the force of gravity on the object. The object is then placed on top of the cliff by moving it along a straight line horizontally. If there is no friction, the stone is raised at a constant speed in order to place it on the cliff. The work done to raise the mass is:

$W_k = Fh$ Where F is applied upward and is equal in magnitude to the force of gravity.

$W_k = mgh$ By Newton's second law.

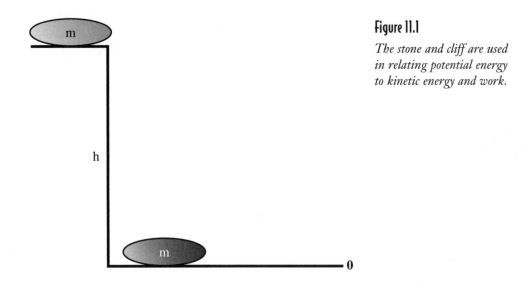

Figure 11.1

The stone and cliff are used in relating potential energy to kinetic energy and work.

A Johnnie's Alert _____

The expression *mgh* is the amount of work done on an object by an applied force upward equal in magnitude to the force of gravity on the object as it is lifted to a height *h* above an arbitrary zero position in a gravitational force field. That means that the expression *mgh* is correct only for objects near the surface of the earth where the acceleration due to gravity is constant.

When *mgh* amount of work is completed, the object is moved horizontally without doing any more work and placed on top of the cliff. You know what happens when the object is pushed off the cliff. (This is when my students would volunteer to have me, the densest object in the classroom, play the role of the one being pushed off the demonstration table to confirm their hypotheses.)

Since you know what would happen, you understand why scientists say that the object has potential energy as a result of having been placed in the gravitational field *h* units above the ground. In fact, the object is said to have potential energy equivalent to the work that was done in placing it at that level in the force field.

At the top of the cliff, $W_k = PE_{top}$ so $PE_{top} = mgh$; remember this quantity is positive because work was done on the object to get it to its new position and its height is now above the zero level. You can see that the object on top of the cliff has zero kinetic energy since it is not in motion. If you shove the object over the side, it will gain kinetic energy and just before it touches the ground the potential energy will be zero. At that point, on the other hand, the kinetic energy is at a maximum. Of course, when it hits the ground it has neither.

It is reasonable to believe that, under ideal conditions, the sum PE + KE = constant. In this case, the constant will be *mgh*. We calculate that sum at two places other than the top and bottom of the cliff to see if that is true.

Suppose that you are able to freeze the object momentarily in space and time the instant it has fallen $\dfrac{h}{2}$ units, which means that it is $\dfrac{h}{2}$ units above the ground:

$-\dfrac{h}{2} = \dfrac{v'^2 - v^2}{-2g}$ Since it is uniformly accelerated motion.

$gh = v'^2$ Because $v = 0$.

So $KE = \dfrac{1}{2}m(gh)$ by substitution and $KE = \dfrac{mgh}{2}$ and $PE = mg\left(\dfrac{h}{2}\right)$ because the object is $\dfrac{h}{2}$ units above ground. Now $KE + PE = \dfrac{mgh}{2} + \dfrac{mgh}{2} = mgh$, the constant that we predicted!

You might try one more position. Suppose that you can freeze the object again when it has fallen $\dfrac{2}{3}h$ units, which means it is $\dfrac{h}{3}$ units above the ground:

$-\dfrac{2}{3}h = \dfrac{v'^2 - v^2}{-2g}$ Because the object has fallen $\dfrac{2}{3}h$ while accelerating at the acceleration of gravity.

$\dfrac{4}{3}gh = v'^2$ Because $v = 0$.

So $KE = \dfrac{1}{2}m\left(\dfrac{4}{3}gh\right) = \dfrac{2}{3}mgh$ and $PE = mg\dfrac{h}{3}$, therefore PE + KE = mgh! Any time that friction is negligible for an object in vertical motion near the surface of the earth, the sum of the kinetic energy and potential energy is constant.

Physics Phun

1. Water at the top of Niagara Falls has potential energy with respect to the pool 167 ft below. How much potential energy does 5.0 ft^3 of water have at the top of the falls? What is the speed of the water just before it touches the pool at the foot of the falls?

The weight density of water is $62.4 \dfrac{lb}{ft^3}$, which means 1.00 ft^3 of water weighs 62.4 lb. Hint: The potential transfers to kinetic energy as the water falls.

2. A 20.0 kg crate slides down an inclined plane that is 3.00 m high and 20.0 m long. If friction is negligible, what is the speed of the crate when it reaches the bottom of the plane? Hint: Use energy considerations in your solution.

Potential Energy of Objects Above the Earth

An object pushed off the top of a cliff is above the earth until it reaches ground zero. What about something that leaves ground zero, goes up, stops, turns around, and returns? You have seen that behavior before. I promised you a solution that is more complete and satisfying after supplying you with a little more information.

Revisit the projectile again from Chapter 7. Figure 7.2 is good enough for you to get an explanation that includes potential energy. Do you remember that kinetic energy seemed to disappear and reappear again? There really are no tricks and no magic to this problem, just some plain old physics.

A High-Flying Potato

Use an initial velocity of 30.0 m/s at an angle of 40.0° again if you want to, because you have done some of the arithmetic already. Of course, you may use any reasonable velocity and angle you choose and find that you get the same result. Recalling the geometric construction you made to find the horizontal and vertical components of velocity, you found $v_{0H} = 23.0 m / s$ and $v_{0V} = 19.3 m / s$. You also found that

$KE = \dfrac{1}{2}(0.114 kg)(30.0 m / s)^2 = 51.3 joules$ for the kinetic energy of the tater when it

was launched. In addition, you found that the kinetic energy had become less by the time the potato reached the top of its arc.

Beginning with that information you can now fill in some of the blanks we had before and appreciate the rest of the story. Start by noting that the potential energy at the instant of launch is zero by using the level of the launch as zero potential energy. You may choose any level you like to represent zero potential energy but usually the person solving the problem will use the lowest level that is convenient and lends itself to an easier solution to the problem.

If the projectile behaves like the stone, you would expect the sum of the kinetic energy and potential energy to be constant. After all, the projectile is still near the surface of the earth where the acceleration due to gravity is constant. Check to see if the sum of the kinetic energy and potential energy at the maximum height is equal to the kinetic energy at launch.

$$KE = \frac{1}{2}(0.114kg)(23.0m/s)^2 = 30.2\,joules$$

$$PE = (0.114kg)(9.80m/s^2)(maximum\ height)^2$$

Hmmm! What is the maximum height of a projectile? You can derive that for yourself if you have forgotten. Maybe you would rather look it up in your notes. No matter how you do it at this time, make sure you can develop any of these ideas you need when you need them. The maximum height is $Y = \frac{v_{0v}^2}{2g}$. For this problem $Y = \frac{(19.3m/s)^2}{2(9.80m/s^2)} = 19.0m.$

That is a high-flying potato! Now you want to check the sum of kinetic energy and potential energy at Y.

$$KE + PE = 30.2\,joules + (0.114kg)(9.80m/s^2)(19.0m) = 30.2\,joules + 21.2\,joules = 51.4\,joules$$

That is within about 0.2% of our initial kinetic energy plus potential energy so that is good enough for that check. You may want to look at KE + PE when the projectile is at $\frac{Y}{2}$ just to make sure.

Instantaneous Velocity of the Potato

You may be convinced already, but you would not want to miss a chance to review some ideas at the same time you are becoming familiar with potential energy. You can develop an expression for $\frac{Y}{2}$ in terms of v'_v, the vertical component of the instantaneous velocity at $\frac{Y}{2}$, and use that along with the horizontal component of the initial velocity, which remains the same throughout the flight, to determine the instantaneous velocity at that point. You can look it up or maybe you have it right at the tip of your tongue. Oh, that is right, I promised that I will show you step by step. If you are ready, here we go!

$\dfrac{Y}{2} = \dfrac{v_V'^2 - v_{0V}^2}{-2g}$ This is the expression you wanted.

$-Yg = v_V'^2 - v_{0V}^2$ By multiplying both members by $-2g$.

$\dfrac{-v_{0V}^2}{2} = v_V'^2 - v_{0V}^2$ By substituting $\dfrac{v_{0V}^2}{2g}$ for Y and then simplifying.

$\dfrac{v_{0V}^2}{2} = v_V'^2 - v_{0V}^2$ By adding v_{0V}^2 to each member.

$v_V' = \pm\sqrt{\dfrac{v_{0V}^2}{2}} = \pm\dfrac{v_{0V}}{\sqrt{2}} = \pm\dfrac{19.3m/s}{\sqrt{2}} = \pm 13.6m/s$

Did you notice that this solution gave us two correct answers? Oh, well, that is all right because we have one value for the trip upward and one for the part when it heads toward the earth. Splat!

The instantaneous velocity at $\dfrac{Y}{2}$ is found by $v_Y = \pm\sqrt{(13.6m/s)^2 + (23.0m/s)^2} = \pm 26.7m/s$.

Why? Well, you know that the vertical velocity decreases as the projectile rises until it becomes zero and the maximum height. The instantaneous velocity is tangent to the path at every instant and has components v_V' and v_{0H} that are mutually perpendicular. That means we can use the Pythagorean theorem to calculate its magnitude. Remember that the length of the hypotenuse of a right triangle is equal to the square root of the sum of the squares of the legs of the triangle.

The magnitude of the instantaneous velocity is equal to the square root of the sum of squares of its components. That is why we found $v_Y = \pm 26.7m/s$, the magnitude of v_V' at height $\dfrac{Y}{2}$, which is half the maximum height the projectile reaches. You can check these ideas at any height you wish by remembering the height at any time it is in the air is given by $y = v_{0V}\tau - \dfrac{1}{2}g\tau^2$.

Newton's Figs

A handheld calculator can save you hours of laborious arithmetic. A graphing calculator is even better to check some of the suggestions I make about the path of a projectile or other objects in motion.

You can get back to your check to see if it works for these two new positions.

$(KE + PE)_{\frac{Y}{2}} = \dfrac{1}{2}mv_Y^2 + mg\left(\dfrac{Y}{2}\right) = \dfrac{1}{2}(0.114kg)(\pm 26.7\ m/s)^2 +$

$(0.114kg)(9.8m/s^2)\left(\dfrac{19.0m}{2}\right) = 51.2\ joules$

This is essentially the same as all other checks so far. Notice that this check covers both places in the trajectory where the height is $\frac{Y}{2}$. When you calculate the kinetic energy, squaring the velocity, whether it is positive or negative, will give you a positive value. You might look at one last check for our purposes: the situation just before the projectile touches the ground.

The velocity is negative because it is going down; that is all right. The potential energy is zero because the height is zero, and you have no problem with that. The magnitude of the instantaneous velocity is the same as when the projectile was launched. Therefore, the kinetic energy is the same as the initial kinetic energy the instant of launch.

Newton's Figs

Review the process of resolving a vector into mutually perpendicular components. Reverse that process either by a graphical solution or by using the Pythagorean theorem to calculate the magnitude of the original vector. If you need the angle associated with the original vector, you must construct the components to scale unless you are a mathematical wizard.

This means that for baseballs, footballs, taters, and spit wads the sum of the kinetic energy and potential energy is constant so long as friction is negligible. I included that disclaimer because you know as well as I do that some projectiles may be launched fast enough that they slow down or even begin to melt because of friction! Some projectiles, if launched high enough and fast enough, will fall around the earth in a circular path!

Orbital Velocity

Do you remember when we compared the weight of an object near the surface of the earth with its weight at any altitude in Chapter 8? Newton's Universal Law of Gravitation was used to make that comparison, and we can use that law to clarify an idea I raise here, the acceleration due to gravity at any altitude. Newton's Universal Law of Gravity enables you to calculate the acceleration due to gravity at any altitude, g_A, by calculating the weight of an object at any altitude.

$W_A = G\frac{m_e m}{R^2}$ Where m_e is the mass of the earth and R is the distance from the center of the earth to the object.

$mg_A = G\frac{m_e m}{R^2}$ By Newton's second law.

$g_A = G\frac{m_e}{R^2}$ By dividing both members by m.

LandSat is a satellite with a polar orbit and is at an altitude of 570 miles. You can calculate the acceleration due to gravity at that altitude.

The radius of the earth is about 3960 miles, so $R = 3960mi + 570mi = 4530mi$. The acceleration due to gravity is:

$$g_A = \frac{6.67 \times 10^{-11} N - m^2 / kg^2 \left(5.96 \times 10^{24} kg\right)}{\left(7.29 \times 10^6 m\right)^2} = 7.48 m / s^2$$

You can see that this value is considerably different than $9.80m / s^2$, the constant value you have used for the acceleration due to gravity near the surface of the earth. If we had a tower high enough—570 mi is pretty high—then we could launch a satellite horizontally at a speed that would cause the satellite to fall around the earth in a circular orbit.

Newton's Figs

The acceleration due to gravity at the altitude of an orbiting satellite is the centripetal acceleration of the satellite as it moves in its nearly circular path at a constant speed.

Actually, as you know, satellites (as well as manned spacecraft) are launched by rockets that lift the payload vertically. When it reaches the peak of its path and is horizontal, other rockets are fired to give the payload the speed it needs to orbit. This is an oversimplified version of what takes place, but you have probably watched all these things take place on television. You can calculate the orbiting velocity of a satellite so that you will know what speed that final stage must provide to the payload. The centripetal force that keeps the satellite in circular orbit is equal to its weight. That can be stated as:

$$f_g = f_c$$

$mg_A = \dfrac{mv^2}{R_A}$ Substituting and using Newton's second law.

$g_A = \dfrac{v^2}{R_A}$ Because the mass remains constant, it is the same in both places.

$v^2 = g_A R_A$ Multiplying both members by R_A.

$v = \sqrt{g_A R_A}$

That means the orbiting speed is equal to the square root of the product of the acceleration due to gravity at that altitude and the distance the satellite is from the center of the earth at that altitude. The orbiting speed of the LandSat is calculated to be:

$$v = \sqrt{\left(7.48m / s^2\right)\left(7.29 \times 10^6 m\right)} = 7.38 \times 10^3 m / s$$

That may not be a familiar figure to you, so you can change it to mi/hr:

$$7.38 \times 10^3 \frac{m}{s} \times \frac{1mi}{1610m} \times \frac{60s}{\min} \times \frac{60\min}{hr} = 16,500mi/hr$$

That figure may be a little more familiar. Usually a commentator will give you a figure of about 17,000 mi/hr for the orbiting velocity of a spacecraft.

Newton's Figs

Once you know the velocity required to keep a satellite in orbit, you can calculate its period. Using that information, you can calculate the mass of the earth. You can also calculate the mass of the earth using the orbiting velocity, $m_e = \frac{R_A v_A^2}{G}$, which you can derive from the statement that the centripetal force on the satellite at any altitude is equal to the gravitational force between the earth and the satellite at the same altitude above the earth.

You can see that this is out of the realm of the projectile launched near the surface of the earth with much smaller velocities. We can account for the kinetic energy and potential energy of the projectile because the acceleration due to gravity remains constant.

Here, as you can see, the acceleration due to gravity gets less and less the greater the distance from the earth. You will need much more mathematics to calculate the potential energy of objects as far away as satellites. You can calculate the kinetic energy of the satellites because you know how to calculate their velocities. You can also use that same information to calculate the period of motion of the satellite. Just to help you to recall how that is done, here is the algebra:

$$g_A = \frac{4\pi^2 R_A}{T^2} \text{ and } T^2 = \frac{4\pi^2 R_A}{g_A} \text{ so } T = 2\pi \sqrt{\frac{R_A}{g_A}}$$

Transfer of Potential Energy

You have actually worked a great deal with this idea already even though I did not call it a "transfer of energy." That should not be confused with the transformation of energy, which is discussed in Chapter 16.

Look at your analysis of the energy of the projectile. It had a maximum amount of kinetic energy initially and no potential energy. If you think about the projectile traveling up to the maximum height, the kinetic energy gets less with height and the potential energy increases. You can think of this as a decrease of kinetic energy and a corresponding increase of potential energy. Note that no energy disappears, as it seemed when you first observed kinetic energy.

The potential energy reaches a maximum value when the projectile is at maximum height, where the kinetic energy is minimum. On the way down, the potential energy decreases and there is a corresponding increase in kinetic energy because the speed of the projectile increases.

When the projectile is about to touch the ground, the potential energy is zero and the kinetic energy is at a maximum value. As you discovered in several examples, the sum of the kinetic energy and potential energy is constant so long as friction is negligible. You will find later that when friction is involved, part of the kinetic energy is used to counter friction and part is transferred to potential energy.

Refer to Figure 11.2 to revisit the simple pendulum, which is another good example of the transfer of kinetic energy to potential energy and back again. Notice that the pendulum is raised to a height h on the right side of the diagram, and at that position it has potential energy but no kinetic energy. In fact, PE = mgh, the work done in lifting the pendulum to height h.

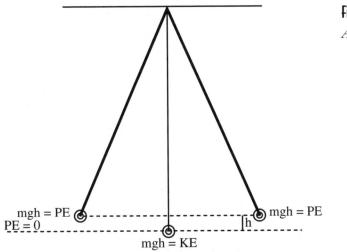

Figure 11.2

An example of energy transfer.

For small angles, the pendulum swings smoothly when released on the right, and as it swings toward the equilibrium position, part of the potential energy is transferred to kinetic energy. When it reaches the equilibrium position, notice that the value of the KE is the same as the value of *mgh*, the amount of potential energy it had at height *h*, because the potential energy is zero at that level.

As it swings through the equilibrium position, it has enough kinetic energy to keep it moving up to a height *h* at the left of the diagram. The pendulum is at rest momentarily at that position with potential energy but no kinetic energy. In fact, all of the kinetic energy it had at the equilibrium position was transferred to the pendulum as potential energy to get it to height *h* at the left of the diagram. You know that as the pendulum swings, the sum of the kinetic energy and potential energy is constant. The constant in this case is the value of the original *mgh*, the amount of potential energy given to the pendulum when it was placed *h* units above the zero potential energy level in the diagram. That is how the pendulum swings. Just like Galileo's chandelier at church.

Hooke's Law

The kinetic energy and potential energy of an object moving in the earth's gravitational field is fairly straightforward. There are other force fields that provide for potential energy and kinetic energy. One other force field familiar to you is that encountered in the stretching of a spring. Hooke's law states that the force required to stretch or compress a spring is directly proportional to the spring's elongation or compression. Each spring has a characteristic constant of proportionality. Hooke's law can be stated symbolically as $F = -kx$, where k is the constant of proportionality, x is the elongation or compression, and F is the force of the spring. Refer to Figure 11.3 as I develop the energy considerations for a spring behaving according to Hooke's law.

> **CAUTION**
>
> **A Johnnie's Alert**
>
> Remember that the force you apply to stretch a spring in Chapter 6 is considered positive because you are doing work upon the spring; that is, you are adding potential energy to the spring. The force the spring applies to you is considered negative from the spring's point of view.

Figure 11.3

A spring stretched by a mass according to Hooke's law used to discuss kinetic energy and potential energy.

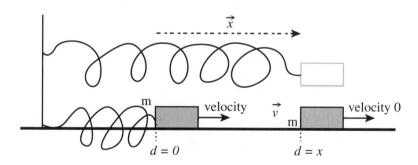

An object with a mass m is on a horizontal frictionless surface and has an initial velocity $\vec{v}$ when it started stretching a spring. After the velocity of the object is reduced to zero, caused by an elongation of the spring by an amount $\vec{x}$, it reverses its direction. As the spring stretches, elastic potential energy is stored in the spring until all of the kinetic energy of mass m has been transferred to the spring. At that instant, the spring starts transferring the elastic potential energy to mass m and continues to do so until the mass has its original kinetic energy back. That happens when the elongation of the spring is again 0.

You should focus on this series of events even though you know that if the mass is free to move it will continue in motion to the left in the diagram. The spring is compressed, storing energy as elastic potential energy until all of the kinetic energy is transferred to the spring as elastic potential energy. The elastic potential energy will return to the mass, and it will regain its original kinetic energy as it returns to its equilibrium position. The mass will oscillate back and forth along the horizontal surface. Remember that friction and the mass of the spring are negligible for the purposes of discussion here.

 **Newton's Figs**

Hooke's law enables you to calculate the force exerted by the spring as simply $F = -kx$. Because of that relationship, the average force is calculated as the initial force plus the final force needed to elongate the spring, divided by two. Average force is not always that easy to calculate.

When mass m first begins to stretch the spring it has traveled a distance 0, so no energy is stored in the spring. The total energy of spring and mass is the kinetic energy of mass m,

$KE = \frac{1}{2}mv^2$. The force applied to the spring by the mass is 0 because the spring has not been elongated.

As the spring stretches, the mass applies a force equal to k times the amount of elongation until all of the kinetic energy has been stored as elastic potential energy. That means the force exerted on the spring is the maximum force, $F = kx$. The force varies from 0 to kx as it is stretched. In order to calculate the work done stretching the spring, you must calculate the average force to cause the elongation x. The average force is given by $F_{avg} = \frac{F_{initial} + F_{final}}{2} = \frac{0 + kx}{2} = \frac{kx}{2}$. The work done to stretch the spring is calculated as $W_k = F_{avg}x = \left(\frac{kx}{2}\right)x = \frac{kx^2}{2}$. The work done to stretch the spring is stored in the spring as elastic potential energy, $PE_{spring} = \frac{1}{2}kx^2$. You now have enough information to account for the transfer of energy from KE of mass to PE of spring and back again:

◆ When the elongation of the spring is zero, KE + PE = $\frac{1}{2}mv^2 + 0 = \frac{1}{2}mv^2$.

◆ When the elongation is x, KE + PE = $0 + \frac{1}{2}kx^2 = \frac{1}{2}mv^2$. That means that the value of the elastic potential energy at the end of the stretching is equal to the value of the kinetic energy the mass had before it started stretching the spring.

♦ When the spring returns to the equilibrium position, the elastic potential energy of the mass is 0, $KE + PE = \frac{1}{2}mv^2 + 0 = \frac{1}{2}kx^2$. This means that all of the elastic potential energy stored in the spring is now returned to the mass as kinetic energy.

♦ Between 0 elongation and an elongation of x, PE + KE = constant. In this case the constant is the kinetic energy the mass had initially.

The process may seem a little complicated at first, but if you run through it a few times it will become clear. You observe the same process of transfer of energy for a spring as we observed for the projectile and the pendulum.

The Least You Need to Know

♦ Calculate the potential energy of an object near the surface of the earth.

♦ Account for kinetic energy and potential energy transferred in the path of a projectile at four different positions.

♦ Use work, kinetic energy, and potential energy to describe the motion of an object on an inclined plane.

♦ Calculate the potential energy stored in a spring.

♦ Explain the energy transfer in the motion of a simple pendulum.

Conservation of Energy

In This Chapter

- ◆ Energy does not just disappear
- ◆ If there is impulse, there is momentum
- ◆ Keeping books on momentum
- ◆ Two billiard balls meet head-on
- ◆ Increasing and decreasing energy

This chapter brings together most of the ideas that have been discussed from the beginning of this book. It should be clear to you by now the importance of becoming familiar with certain ideas before tackling others. You find that motion and force lead to an understanding of work, kinetic energy, and potential energy. The inclusion of impulse and momentum in the discussion enables you to develop even more details in this intricate mosaic of concepts called physics.

I know that you can function just fine without knowing how nature works. However, it puts you at a disadvantage to have no fundamental under-standing about the physical laws that govern the interactions of matter and energy. You now know that if you are driving your car at 60 mph instead of 30 mph, and you hit a tree; there will be four times as much energy doing damage to you and the tree, not just the double amount that many think. You know this because of your work with $\frac{1}{2}mv^2$.

In this chapter I will give you even more evidence that there are people who spend their lives in an effort to give meaning to all of these strange-sounding terms and processes. Isn't that scary? It is, and wonderful as well to learn that men and women have given us explanations for physical phenomena that we use to our advantage and take for granted in our daily lives.

Early man thought that fire was a wondrous thing, and it is, especially if it is controlled. Imagine what a person just 150 years ago would think about a video camera that plays back recordings. Our scientists have spoiled us, but we must exert some effort to find out just how much.

There is another property of matter that must be discussed in order for you to understand fully what occurs if your car happens to hit the tree that I talked about earlier. Let's get started so that you will have an even better understanding of how your physical world works.

Where Does Energy Go?

So far our work with mechanical energy indicates that kinetic energy and potential energy do not go anywhere. That is, as long as we do not have to take friction into account. Several examples provided you with the information you needed to show that the sum of potential energy and kinetic energy is constant in an isolated system.

In our discussion of the path of a projectile, you found that as the projectile left the launching level and climbed upward the kinetic energy became less and the potential energy increased. You can say that it decreased in kinetic energy and increased in potential energy, always retaining the same total energy. When it reaches the maximum height, it has minimum kinetic energy and maximum potential energy. On its way down, the projectile increases in kinetic energy and decreases in potential energy. Just before it touches the ground, or reaches the launching level, the potential becomes zero and the kinetic energy is at its maximum value. There is no decrease in the total mechanical energy under ideal conditions. You can say the mechanical energy is conserved, that is, there is no gain or loss of mechanical energy.

Newton's Figs

You have the kinetic energy and potential energy as added tools for solving problems involving work and machines. Use those ideas when you need more information about motion along an inclined plane or when something is to be raised straight up by an inclined plane or lever.

Earlier there was a hint that friction may introduce a different result. That means that if the moving object encounters friction, some energy with other properties and a different name appears. There could be thermal energy as there is when a piece of wood is rubbed with another piece of wood. If the thermal energy is large enough to increase the temperature of the wood to its kindling temperature, then a fire starts. Often, as a result, there is energy with new properties we call light. While the pieces of wood are being rubbed together, there is a distinctive sound created that can be quite irritating, as sound energy appears.

Impulse and Momentum

You have probably had some practice tossing a ball into the air and hitting it with a bat. Suppose that you use a baseball bat and toss a baseball into the air. Also suppose that you are good enough to hit it. Your result would be a quick way to launch a projectile.

What do you suppose would happen if you tossed a 4 kg shot up and hit it with a bat? The bat would most likely break, but the shot would probably move a bit as well. The thing that you would notice in both cases is that the bat would apply a force for a short time to the baseball and the shot. Instead of using a bat, launch the shot by "putting" it like an athlete does with one heave (hence the event name *shot put*)—then do the same thing with the baseball.

You probably pushed each ball into the air with the same force and held each for about the same amount of time. The baseball had a much larger change in velocity than did the shot because of their different masses. In order to give the shot the same change in velocity as the baseball, you would either have to apply a much larger force for the same amount of time or the same force for a longer period of time. This experiment is somewhat unrealistic because the shot weighs about ten pounds. I used these two objects because of the obvious difference in the way they behave when you push them both with the same force for the same amount of time.

In physics, the idea that is involved here is called *impulse*. The bat gave both balls the same impulse and each behaved in a much different way. Impulse is a vector quantity and is calculated by multiplying the applied force by the time interval during which it is applied.

Plain English

Impulse is a physical quantity that results from a force being applied to a body for a certain amount of time.

In symbols, $\vec{I} = \vec{F}\Delta t$ where the direction of the impulse is the same direction as the force applied. Yes indeed, you know that the next thing we should do is check the units of impulse in each system. You use the symbolic representation of impulse and find that impulse is measured in N-s in the MKS system, dyne-s in the CGS system, and lb-s in the FPS system. Those units may be expressed as s-N, s-dyne, and s-ft.

Newton's Figs

A change in momentum usually implies a change in velocity, but keep in mind that the mass can change also as in the case of a rocket launched from the earth. Not only is the fuel being used up, making the mass of the rocket less, but also the fuel tanks themselves are jettisoned after a short period of time.

Plain English

Momentum is a physical vector quantity that has its magnitude determined by the product of the mass and velocity of the object being described.

A Johnnie's Alert

Any time an object experiences an impulse, the object gives an impulse to the source at the same time. That is, the bat gives the ball an impulse and the ball gives the bat an equal and opposite impulse.

You know that when an object receives an impulse, it undergoes a change in velocity. The magnitude of the change in velocity depends upon the mass of the object. The more massive the body, the smaller the change in velocity it will realize from a given impulse. The physical property that is involved here is called *momentum.*

Momentum is a vector quantity, and its magnitude depends on the mass and change in velocity of the object being described. Its direction is in the direction of the velocity vector. Momentum is important enough to have a special symbol, $\vec{P}$, assigned to it. Momentum is calculated as $\vec{P} = m\,\vec{v}$. The units of momentum are $\dfrac{kg - m}{s}$ in the MKS system, $\dfrac{g - cm}{s}$ in the CGS system, and $\dfrac{slug - ft}{s}$ in the FPS system. You have probably already noticed that these units are a disguise of N-s, dyne-s, and ft-s. The reason is the relationship between impulse and momentum.

Remember how the same impulse gives a different change in velocity to two objects of different mass? When an object is given an impulse, it undergoes a change in momentum. You see momentum involves both mass and change in velocity. The mathematical statement of this relationship is $\vec{F}\Delta t = \Delta\left(m\,\vec{v} \right)$; in this book you will consider changes in momentum of objects that have only a change in velocity and not a change in mass.

Understanding the Relationship Between Impulse and Momentum

Our first look at the relationship between impulse and momentum will involve expressions modeled after $\vec{F}\Delta t = m\Delta\,\vec{v} = m\left(\vec{v'} - \vec{v} \right) = m\vec{v'} - m\,\vec{v}$ and you should focus on the magnitude of these quantities unless it is explicit that they are involved in motion that is not along a straight line. Now you can see why the units of impulse are the same as the units of momentum even though the units for impulse are N-s and for momentum $\frac{kg-m}{s}$ in the MKS system, that is, $N-s=\frac{kg-m}{s^2}\times s=\frac{kg-m}{s}$. You can do the same analysis of the other systems of measurement to find that the units for impulse are the same as the units for momentum.

A closer look at this new relationship will reveal to you why it is thought that this is the relationship Newton originally used as his second law. That is, $\vec{F}\Delta t=m\vec{v}'-m\vec{v}$ can be stated $\vec{F}=\frac{m\vec{v}'-m\vec{v}}{\Delta t}=m\left(\frac{\vec{v}'-\vec{v}}{\Delta t}\right)=m\vec{a}$. His third law follows from the line of reasoning summarized here.

If the bat gives the ball an impulse, the ball gives the bat an impulse—you can feel it. Therefore, when the bat applies a force on the ball, the ball applies a force on the bat, equal in magnitude, but in the opposite direction, for the same period of time.

Suppose you have two objects and the first gives the second one an impulse, then the second will give the first an impulse also. That means that $\vec{I_1}=-\vec{I_2}$ and $\vec{F_1}\Delta t=-\vec{F_2}\Delta t$, since the time of the impulse is the same $\vec{F_1}=-\vec{F_2}$. The forces are equal in magnitude but opposite in direction, so for every action there is an equal and opposite reaction. When Newton used the words action and reaction, he meant force and reactive force.

Physics Phun

1. A projectile weighs 151 lb. What net force is required to accelerate the projectile at a rate of $925\frac{ft}{s^2}$?

2. (a) If a force of 5.00 N acts on an object for 3.00 s, calculate the change in momentum of the object. (b) If the object has a mass of 5.00 kg, calculate its change in velocity. (c) If its original velocity were 1.0 m/s, what would be its final velocity?

3. A force of 35.6 N acts on a body for 0.250 seconds. What is the impulse that acts on the body?

Conservation of Momentum

Suppose that you have two steel balls on a table, and you roll one of the balls, m_1, toward the other ball, m_2, that is at rest initially. Eventually they collide. You have done that many times before, probably either with marbles or billiard balls. You know the balls fly off in different directions if there is a glancing blow. Other things can happen, like the first one stopping, and the other going off in the direction of travel of the ball that was rolled initially.

Later you will get a chance to do a complete job of describing the last case, but for now concentrate on the general idea of what is going on. You know that any two objects that collide, or interact, apply impulses one to the other. You can summarize what happens as:

$$\vec{F}_1 \Delta t = -\vec{F}_2 \Delta t \quad \text{Their impulses are equal in magnitude and opposite in direction.}$$

$$\Delta \vec{P}_1 = -\Delta \vec{P}_2 \quad \text{Substituting the change in momentum for impulse.}$$

$$(m_1 v_1' - m_1 v_1) = -(m_2 v_2' - m_2 v_2) \quad \text{Substituting for change in momentum.}$$

$$m_1 v_1' - m_1 v_1 = m_2 v_2 - m_2 v_2' \quad \text{Simplifying.}$$

$$m_1 v_1' + m_2 v_2' = m_2 v_2 + m_1 v_1 \quad \text{Adding to both members } m_1 v_1 + m_2 v_2'.$$

A Johnnie's Alert

When you read a problem in physics, record the given information first and make sure all data are in the same system of measurement. If different systems are employed, use the techniques of unit analysis to express everything in the same system before you begin your solution.

Newton's Figs

An algebraic solution to a physics problem serves many purposes, the main one being that it provides you with a template you can use to find as many numerical solutions as you like.

Notice that the last equation has in its left member the momentum of ball one and ball two after the interaction and the right member has the momentum of ball one and ball two before the interaction. Another way of stating the same thing is the sum of the momenta (plural of momentum) after the interaction is equivalent to the sum of the momenta before the interaction. That is, the total momentum before the interaction is the same as the total momentum after the interaction; the total momentum is constant. There is no change in the total momentum.

This is a statement of the law of the conservation of momentum. This law applies even if the interacting bodies stick together after the interaction; momentum is still conserved. In an interaction of two bodies, total momentum is neither gained nor lost. You have probably seen a case like this when freight cars are being joined to form a train. One car is usually pushed toward another; the two interact and lock together then move off together at a slower speed than the first car had initially.

If two automobiles interact at a high speed, both cars crunch, lock together, and the mess moves off in a different direction than either was traveling initially, then momentum is conserved. Momentum is conserved for billiard balls, freight trains, crashing cars, and other objects. A few examples of interactions may help you to tie these ideas together.

Example 1: Conserving Momentum

A 50.0 kg girl rides on a 15.0 kg skateboard that is moving in a straight line at a constant velocity of 3.0 m/s. The girl jumps off the skateboard. What is the change in velocity of the skateboard for each of the following cases?

(a) She jumps off at three times the original velocity of the skateboard.

(b) She jumps off with the same velocity as the skateboard had originally.

(c) She jumps off not moving with respect to the ground.

The outline of the solution is the same as used before but I review them here to emphasize the nature of thinking involved in the solution of a physics problem: (1) List all information given and implied, and the information you are to solve for. (2) Assign appropriate algebraic symbols to all information in (1). (3) Solve the problem algebraically first, then substitute numerical values to identify a numerical solution. Sometimes a diagram that is properly labeled will help you to devise a strategy for a solution. Vector diagrams can provide a numerical solution by either a graphical construction or a calculation involving geometrical techniques.

Solution:

$$m_g = 50.0 kg$$

$$m_s = 15.0 kg \qquad (a)\ v'_g = 3v_s$$

$$v_g = v_s = 3.0 m/s \qquad (b)\ v'_g = v_s$$

$$\Delta v_s = v'_s - v_s = ? \qquad (c)\ v'_g = 0$$

$$v_g = 3.0 m/s$$

The information given and the information you are to find suggests conservation of momentum considerations.

The sum of the momenta after the interaction is equal to the sum of the momenta before the interaction.

$$m_g v_g + m_s v_s = m_g v'_g + m_s v'_s \quad \text{Conservation of momentum.}$$

$$m_g v_g - m_g v'_g = m_s v'_s - m_s v_s \quad \text{Adding to each member } \left(-m_g v'_g - m_s v_s\right).$$

$$m_g \left(v_g - v'_g\right) = m_s \left(v'_s - v_s\right) \quad \text{Simplifying by factoring common factors in each member.}$$

$$v'_s - v_s = \frac{m_g}{m_s}\left(v_g - v'_g\right) \quad \text{Solving for } v'_s - v_s \text{ by dividing both members by } m_s.$$

$$\Delta v_s = \frac{m_g}{m_s}\left(v_g - v'_g\right) = \frac{m_g}{m_s}\left(v_s - v'_g\right) \quad \text{Substituting } \Delta v_s \text{ for } \left(v'_s - v_s\right) \text{ and } v_s \text{ for } v_g$$

because the velocity of the girl is the same as the velocity of the skateboard before the interaction.

This completes the algebraic solution. And now for numerical solutions:

(a) $\Delta v_s = \dfrac{50.0kg}{15.0kg}(3.0m/s - 3(3.0m/s)) = -2.0 \times 10m/s$

(b) $\Delta v_s = \dfrac{50.0kg}{15.0kg}(3.0m/s - 3.0m/s) = 0$

(c) $\Delta v_s = \dfrac{50.0kg}{15.0kg}(3.0m/s - 0) = 1.0 \times 10m/s$

Do these final results make sense? This is a question that should occur to you any time you solve a problem. In this case, in (a), the girl jumped off the front of the skateboard, sending it in the opposite direction. In (b), the girl jumped off the side in order to have no change in velocity. In (c), the girl jumped off the back of the skateboard, sending it ahead faster than before she jumped. You can also find the numerical sum of momenta before the interaction and compare that to the numerical sum of momenta after the interaction. If the problem is solved correctly, the sums should be the same. Are they the same?

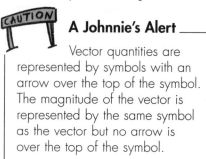

A Johnnie's Alert

Vector quantities are represented by symbols with an arrow over the top of the symbol. The magnitude of the vector is represented by the same symbol as the vector but no arrow is over the top of the symbol.

Example 2: A Total Momentum of Zero

An object at rest is torn into three pieces by an explosion. Two pieces fly off along paths that are perpendicular to each other. One of the two pieces has a momentum of 28.0 kg-m/s and the other has a momentum of 15.0 kg-m/s. What is the mass of the third piece if its speed is 20.0 m/s?

Solution:

Use the same outline as before. In this case, a scale drawing is required because the vector nature of momentum is implied in the statement of the problem when you are given that two pieces fly off along paths that are perpendicular to each other. Refer to Figure 12.1 for the vector solution.

$$\vec{P_1} = 28.0 \frac{kg - m}{s} E$$

$$\vec{P_2} = 15.0 \frac{kg - m}{s} N$$

$$m_3 = ?$$

$$v'_3 = 20.0 m/s$$

$$v_1 = v_2 = v_3 = 0$$

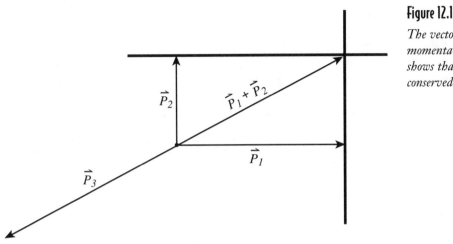

Figure 12.1

The vector diagram of momenta for an explosion shows that momentum is conserved in an interaction.

The sum of the momenta before the interaction, the explosion, must be equal to the sum of the momenta after the interaction. The sum of the momenta before the interaction is zero because the object is at rest. That means the sum of the momenta after the interaction must be zero. The sum of the momenta of two pieces after the interaction is $\vec{P_1} + \vec{P_2} = 32.0 \dfrac{kg - m}{s} E28° N$.

In order to have a sum of zero, the third piece must have a momentum equal in magnitude to the sum of the first two but opposite in direction, $\vec{P_3} = 32.0 \dfrac{kg - m}{s} W28° S$.

That means that $P_3 = m_3 v_3$ and $m_3 = \dfrac{P_3}{v_3}$, so $m_3 = \dfrac{32.0 \dfrac{kg - m}{s}}{20.0 m / s} = 1.6 kg$. Notice that my final answer has two significant figures even though the data is expressed in three significant figures. Since I constructed a vector diagram, my measurements with the ruler had only two significant figures.

We cannot calculate more accurately than we can measure, so my measurement limited the number of significant figures in my answer. I calculated the answer also, and found the mass to be 1.59 kg. The two answers are the same, considering methods used in each case.

Newton's Figs

If you use a scale drawing as part of your solution to a vector problem, the number of significant figures used in your calculations will be limited by the precision of the measuring instruments used in constructing the solution.

Example 3: Momentum of a Combination

A freight car having a mass of 14,200 kg and traveling at 15.0 m/s bumps into an identical car that has the same mass as the first one, initially at rest on a straight and level railroad track. If they couple and travel away together, what is the speed of the combination?

Solution:

The given data suggests that you apply the law of conservation of momentum. Use the same outline for your solution that you have used before:

$$m_1 = 14,200kg$$

$$v_1 = 15.0m/s$$

$$m_2 = 14,200kg$$

$$v_2 = 0$$

$$v_1' = v_2' = ?$$

You know that the sum of momenta before the interaction is equal to the sum of momenta after the interaction.

$m_1v_1 + m_2v_2 = (m_1 + m_2)v_1'$ Conservation of momentum reflecting that there is one mass after the interaction.

$v_1' = \dfrac{m_1v_1 + m_2v_2}{m_1 + m_2}$ Dividing each member by the mass of the total mess.

$v_1' = \dfrac{m_1}{m_1 + m_2}(v_1 + v_2)$ Since $m_1 = m_2$. The algebraic solution is complete.

You can substitute numerical values to find the numerical solution.

$$v' = \frac{1}{2}(15.0m/s + 0) = 7.5m/s.$$

You now have several examples you may use as models for solving momentum problems. The important thing is that you develop your own strategy and follow your own plan to arrive at a solution that helps you feel confident. A lot of practice will help … so here goes.

A Johnnie's Alert

CAUTION

Models for solving problems in this book are just that. They should not be a pattern for constructing solutions. You should use the general outline for solving problems given in this book to create your own solutions.

Physics Phun

1. A person threw a 2.50 kg object horizontally from a canoe with a speed of 15.0 m/s. The mass of the person is 50.0 kg and the mass of the canoe is 102 kg. What is the velocity of the canoe after throwing out the object if the canoe were originally at rest? The 2.50 kg object was not part of the mass of the canoe.

2. A freight car with a mass of 12,200 kg traveling at 12.0 m/s strikes another freight car that was originally at rest. The cars stick together and move away at a constant speed of 5.00 m/s. Calculate the mass of the second car.

3. A bullet of mass 10.0 g leaves the muzzle of a stationary gun of mass 5.0 kg with a velocity of 355 m/s. What is the velocity of the gun as a result of the bullet leaving the muzzle? The shooter experiences the velocity of the gun as a "kick," which is sometimes referred to as the recoil velocity of the gun. Hint: Watch out for units.

The solution to the third problem in the last set gives you a hint to the explosion of another myth. Some people think that a rocket throws hot gases out of the end of the rocket and pushes on the earth to lift off into space. You know that a tiny gas particle leaving the rocket gives the rocket an impulse, causing a change in momentum in the opposite direction. When all of the millions of tiny particles are fired out the end of the rocket, a tremendous change in momentum is given to the rocket in the opposite direction. The same thing happens when the rocket changes direction in space, but usually a burst of gas in one direction will give the rocket a change in momentum in the opposite direction because of the impulse of the gas.

Head-On Collisions

You have looked at several incidences of objects interacting with each other. I have avoided talking about objects colliding because often it is thought that the objects must touch in order to collide. However, in physics, objects can collide without touching; two objects with the same magnetic poles or the same electrical charge never touch, but they can certainly interact as you have probably witnessed. In this book, I generally use the term interaction to make it clear that collisions with and without physical touching are included in the discussion.

The interaction of two billiard balls will be the vehicle I use to discuss head-on collisions. There are some ideas needed in this discussion that are treated first. The collision of two billiard balls or two steel balls is much different than the collision of two mud balls or two balls of putty. As we consider these differences, we will explode another myth.

You have probably made mud balls at some time in your life; you probably even threw them at something like a wall. When a mud ball hits an object, it tends to splatter. It becomes deformed and never returns to the shape it had before the collision. A ball of putty will do the same thing. A car tends to behave in the same way when it runs into a tree. It becomes distorted and remains that way until you pay a lot of money to get it smoothed out.

The mud ball or ball of putty had to be close enough to the wall to experience the impulse of the wall. The fact that the mud ball splatters means that there are forces within the ball causing parts of the ball to move over other parts and remain displaced. The force of the interaction depends on something other than the distance between the ball and the wall.

Newton's Figs

When two objects interact by touching, the collision is often referred to as a hard sphere collision. The important thing to keep in mind is that objects can collide without actually touching. Two billiard balls collide by touching in a hard sphere collision.

The steel balls hanging from strings bounce off each other and do not remain distorted. They are distorted so little that you would have to be an atom-sized person to notice. The same thing is true of billiard balls. The force of the interaction is experienced as soon as the balls touch. The force of the interaction depends only on the distance between the balls. There is no interaction when they are not touching, but as soon as they are close enough together … they experience the impulse of each other.

Charged objects are the same way. Charged objects do not affect each other until they are close enough to be affected by another charged object. The force of the interaction depends on the distance between the charged bodies. If you think about why the balls in the first example do not go through each other, you realize that each ball's outermost electrons are repelling the other ball's outermost electrons. So when the balls get too close, the forces of repulsion push them apart. They have hit and bounced off each other.

A Johnnie's Alert

A rubber band or a rubber ball is not a very elastic object. However, such objects are more elastic than a mud ball. The closest thing we can come to as a model for a perfectly elastic object is a billiard ball or an ivory ball.

Magnetic poles behave the same way. You have probably brought one north pole of a magnet close to the north pole of another and felt the repulsive force pushing on the magnet you hold. The force of the interaction depends only on the separation of the magnets, charged bodies, billiard balls, or steel balls. There are no other forces involved in the interaction like the attraction of the wall for the mud that sticks to it after a mud ball-wall interaction. The idea that is involved here is elasticity.

Understanding Elasticity

You have probably always thought that a rubber band was highly elastic. That is just not so. If an object interacts with another object in such a way that the force of interaction depends only on the separation of the objects, the objects are elastic. Of course, the billiard balls and steel balls are not perfectly elastic; that kind of an object is ideal. A billiard ball is highly elastic whereas a mud ball is highly inelastic because:

◆ A highly elastic object may be slightly deformed momentarily but is immediately restored to its original shape.

◆ An inelastic object may be permanently deformed and never returns to its original shape.

A rubber band is more on the end of the scale with the mud ball; the billiard ball is on the other end of the scale. You can stretch a rubber band and lay it on a table and see it slowly move toward its original shape. After a bit of stretching, it remains stretched out considerably from its original shape. Oops! There goes another rubber band myth!

If we have two billiard balls interact with each other, they behave in a lot of ways depending on the force of interaction. They can fly off at different angles, or in some cases bounce straight back or stop. A little later we take a closer look at the behavior of two billiard balls interacting along a straight line. The reason this business of elasticity is important is this: Kinetic energy is conserved in elastic interactions. It is not conserved in inelastic interactions. Momentum is conserved in all interactions, but not kinetic energy.

You know that there is no perfectly elastic interaction, so we are talking about ideal situations. Billiard ball to billiard ball interaction is about as close to an elastic interaction as we can get. You know as well as I do that if the billiard ball is distorted at all, even though it flies back to it original shape, some kinetic energy is changed to thermal energy through the action of friction. There is a click when they hit so some kinetic energy is changed to sound. The amount of kinetic energy lost due to heat or sound is negligible compared to the energy of the system of interacting bodies under consideration. So when we say that the interaction is elastic, you know that it is highly elastic and not perfectly elastic. You may treat it as a perfectly elastic interaction in your analysis but keep these other things in mind for future considerations.

A Johnnie's Alert

Even though a small amount of kinetic energy may be transferred as sound or thermal energies in an interaction, the interaction may still be treated as elastic because these transfers are negligible, in most cases too small to measure.

Understanding Inelasticity

You should have an example of the type of thing that we are discussing, so look back at the last set of problems. The second problem in that set is about two freight cars colliding. I found the mass of the second freight car to be 17,100 kg. It is instructive to calculate the kinetic energy of both cars before and after the interaction.

$$m_{c1} = 12,200kg$$

$$m_{c2} = 17,100kg$$

$$v_{c1} = 12.0m/s$$

$$v_{c1} = v_{c2} = 5.00m/s$$

$$KE_{before} = \frac{1}{2}(12,200kg)(12.0m/s)^2 = 878,00\,joules$$

$$KE_{after} = \frac{1}{2}(12,200kg + 17,100kg)(5.00m/s)^2 = 366,000\,joules$$

Newton's Figs

Calculate the kinetic energy before an interaction and after the interaction has taken place to check to see if the interaction is elastic. If there is a large difference in the two figures, then the interaction is inelastic; momentum is conserved but kinetic energy is not.

You can see for yourself that kinetic energy is not conserved in an inelastic interaction; in fact about 58% of the kinetic energy is not accounted for. You thought the interaction was elastic? The trains stuck together in much the same way that two mud balls stick together when they interact.

A better example yet of an inelastic collision would be one between a ball of putty and a billiard ball. They stick together because the force of the interaction depends on something other than the separation of the objects.

Now suppose we have two billiard balls arranged so that one ball, m_2, is originally at rest and the other one, m_1, is in motion with a constant velocity, v_1, and causes a head-on collision. It is instructive to solve algebraically for the velocities of the two balls after the interaction in terms of their masses and their velocities before the interaction. Since this interaction is elastic, there is no change in the total kinetic energy or total momentum. The two balls can have such changes in their individual kinetic energies and momenta.

Let $-\Delta KE_1$ represent the amount of kinetic energy released by ball one and ΔKE_2 represent the amount of kinetic energy absorbed by ball two. Similarly, let $-\Delta P_1$ represent the amount of momentum released by ball one and ΔP_2 represent the amount of momentum absorbed by ball two. The symbols used are $m_1, m_2, v_1, v_1', v_2 = 0, v_2'$ and we are to find v_1' and v_2' in terms of m_1, m_2, and v_1.

You will have several equations in this derivation so I will number some of them in order to make quick reference to them in the solution.

$$-\Delta KE_1 = \Delta KE_2$$

$$-\left(\frac{1}{2}m_1v_1'^2 - \frac{1}{2}m_1v_1^2\right) = \left(\frac{1}{2}m_2v_2'^2 - \frac{1}{2}m_2(0)\right) = \frac{1}{2}m_2v_2'^2$$

$$-m_1\left(v_1'^2 - v_1^2\right) = m_2v_2'^2$$

$$-m_1\left(v_1' + v_1\right)\left(v_1' - v_1\right) = m_2v_2'^2 \quad \text{Factoring a difference of squares.}$$

I. $\quad -\left(m_1v_1' - m_1v_1\right)\left(v_1' + v_1\right) = m_2v_2'\left(v_2'\right)$

$$-\Delta P_1 = \Delta P_2$$

II. $\quad -\left(m_1v_1' - m_1v_1\right) = m_2v_2'$

Divide equation I by equation II and you find:

III. $\quad v_1' + v_1 = v_2'$

Solve equation II for v_2' to get:

IV. $\quad v_2' = \dfrac{m_1v_1 - m_1v_1'}{m_2}$

Substitute IV in III and solve for v_1':

$$v_1' + v_1 = \frac{m_1v_1 - m_1v_1'}{m_2}$$

$$m_2v_1' + m_2v_1 = m_1v_1 - m_1v_1' \quad \text{Multiplying both members by } m_2.$$

$$\left(m_1 + m_2\right)v_1' = \left(m_1 - m_2\right)v_1 \quad \text{Adding } m_1v_1' - m_2v_1 \text{ to both members then simplifying.}$$

V. $\quad v_1' = \dfrac{m_1 - m_2}{m_1 + m_2}v_1$

Substitute V into III and solve for $\dfrac{m_1 - m_2}{m_1 + m_2}v_1 + v_1 = v_2'$:

$$\frac{m_1 - m_2}{m_1 + m_2}v_1 + v_1 = v_2'$$

$$v_1\left(\frac{m_1 - m_2}{m_1 + m_2} + 1\right) = v_2'$$

$$v_1\left(\frac{m_1 - m_2 + m_1 + m_2}{m_1 + m_2}\right) = v_2'$$

$$v_2' = \frac{2m_1}{m_1 + m_2}v_1$$

CAUTION

A Johnnie's Alert

Use symbols that represent physical quantities as shorthand to the construction of a solution to a physics problem. Do not let the symbol distract you from the thought process you use to construct a solution.

You have a solution to the problem of a linear head-on collision that gives the final velocities of the two objects in terms of their initial velocities and their masses. Even though this solution involves only algebra, it is somewhat complicated and deserves your review until you can carry out the solution on your own.

Energy Transfer

You may wonder what the big deal is about a solution to two billiard balls running into each other. I hope you are that skeptical about this situation. It is my purpose to help you to see all of the phenomena we have considered in this chapter as part of one complex picture.

First, you can think of the collision of these two objects as being much different than the collision of two cars. You can make calculations to show that even though momentum is conserved in a car crash, kinetic energy is not conserved. Secondly, you have knowledge of elastic and inelastic collisions that will guide your thinking in general about objects interacting with each other. A third thing is that you can account for the transfer of kinetic energy from one elastic object to another when they collide.

There are many other things that you can think of with the background provided here. What about the loud noise of an inelastic interaction? You know that sometimes there is enough thermal energy generated in an inelastic interaction to cause an explosion. You have a keener sense of the collisions of objects in your everyday world, including the home run from the crack of the bat of a star baseball player.

CAUTION

A Johnnie's Alert

The nature of science dictates that there is no such thing as "the right solution." There are many correct solutions, and it is important that you construct your own solution using the relevant physical principles and an answer that makes sense to you. The solution is the path that you take to your answer. Test your answer, as has been modeled for you in example solutions, to see if it makes sense, and if the units of measurement are correct for your answer.

You have a great deal of information from your everyday life about elastic collisions. You have seen billiard balls collide and understand their behavior in general. Look at your solution to the last problem to see what specific information you have at your disposal.

The two expressions for the velocities of the objects after collision are: $v_1' = \dfrac{m_1 - m_2}{m_1 + m_2} v_1$ and $v_2' = \dfrac{2m_1}{m_1 + m_2} v_1$.

Suppose that $m_1 = m_2$, then $v_1' = 0$ and $v_2' = v_1$! Did you expect that? Two elastic objects with the same mass in a head-on collision will have the object originally in motion stop after the collision and the object originally at rest will take off with the original velocity of the first object. The transfer of energy is remarkable; the first object transfers all of its original kinetic energy to the second object. This discussion contributes to the wealth of information you have when you consider momentum and kinetic energy in collisions of objects both elastic and inelastic.

The Least You Need to Know

♦ Use the relationship between impulse and momentum.

♦ Recognize the units of measurement of impulse and momentum.

♦ Apply the law of conservation of momentum.

♦ Identify an interaction as elastic or inelastic.

♦ Construct solutions to elastic collisions using the conservation of momentum and the conservation of kinetic energy.

Part 4

What Happens When Small Things Move?

Glancing through this part you see several things that are familiar to you in your daily activities. They are a result of small particles in motion. At least the model used in this theory assumes that the small particles in motion can explain ideas like states of matter, pressure, heat, and sound. The funky little particles may be ions, atoms, or molecules. I refer to them as particles, the small building blocks of matter in all states. The theory is called the kinetic theory of matter, which should suggest some familiar ideas to you: particles in motion obeying Newton's laws of motion.

13

Solids, Liquids, and Gases

In This Chapter

- ◆ Fillings for teeth

- ◆ Your drinking water

- ◆ The air you breathe

- ◆ Particle speed makes the difference

- ◆ A model for a gas

The world we live in is filled with many awesome sights that are dominated by solids, liquids, and gases. The huge waterfall from the top of a mountain in Oregon appears to come from solid rock. As the water falls, a fine mist is formed displaying a rainbow-like halo when the light is just right. In this chapter, you will find that these forms of matter are alike in some ways and very different in others.

The explanations are based on models, as is often the case in science. The models may be good ones or may have shortcomings, but they provide a way of summarizing observations and making predictions that may be checked by experiment. Keep a sharp eye on these explanations and use your keen mind to consider them.

Solids

You know that the matter making up the dam that is wide at the bottom is an example of a solid. In solids, the particles are thought to vibrate randomly about a point of equilibrium much like the simple pendulum or the vibrations of the spring in Figure 11.3. The amplitude of the movement of the particles is quite small as you might expect if you think of tiny springs attached to each particle. The particles are about 0.40 nanometers apart and vibrate with periods of about 10^{-13} s.

The *cohesive forces* between the particles are the causes of the vibrating motion. However, since the distances moved by the particles are small, that force is strong enough to cause the particles to stay together and hold a definite shape for the solid. There are also *adhesive forces* involved when solids of different kinds come in contact. For example, when you accidentally scrub your shoe against the wall some of the finish on your shoe adheres to the wall in an embarrassing smudge.

Even though the motion of the particles is somewhat restricted, some particles can escape the surface of one solid and enter the surface of an adjacent solid.

Plain English

Cohesive forces are the attractive forces between particles of the same kind. **Adhesive forces** are forces of attraction between particles of different kinds of matter. **Diffusion** is the movement of particles of one kind of matter into the empty space of a different kind of matter because of the random motion of the particles. **Density** is the amount of matter in a unit volume. Since matter is measured in two different ways, there are two types of density. Mass density is the amount of mass in a unit volume of matter and weight density is the amount of weight in a unit volume of matter. The unit measure of mass density might be kg/m3, and the unit measure of weight density might be N/m3. **Specific gravity** of a substance is the ratio of the density of the substance to the density of water. There is no unit measure for specific gravity since the two measurements cancel.

For example, if you lay a bar of gold on a bar of lead, you find after a long period of time some gold particles in the lead and some lead particles in the gold. That process is called *diffusion*. Diffusion in solids occurs slowly; a flood of particles will not diffuse as can occur in other states of matter. Since the particles cause the solid to have a definite shape, the particles also have a definite volume. That means that the solid has mass and inertia. If it is within a gravitational field it will also have a defined weight. Because a solid has definite weight and mass in a definite volume, it has a definite *density*.

Since density is a general property of all types of matter, now is a fine time to discuss it in a little more detail. Density is a derived quantity and it is a scalar quantity. I use these symbols to represent weight density and mass density: ρ_w for weight density and ρ_m for mass density. The definition of each then is: $\rho_w = \dfrac{w}{v}$ and $\rho_m = \dfrac{m}{v}$. The relationship between the two follows from Newton's second law: $\rho_w = \dfrac{w}{v} = \dfrac{mg}{v} = g\rho_m$.

A quantity that is closely related to density is *specific gravity*. In symbols, specific gravity is expressed as: $spgr = \dfrac{\rho_{ws}}{\rho_{ww}} = \dfrac{\rho_{ms}}{\rho mw}$. These ideas will serve you very well in science and calculus as well as in the development of ideas in this part of the book.

You should note that weight density, mass density, and specific gravity are all scalar quantities and are derived quantities. Specific gravity has no units of measurement since it is a ratio of two quantities having the same units of measurement. The units of weight density are: $\dfrac{N}{m^3}$ in the MKS system, $\dfrac{dynes}{cm^3}$ in the CGS system, and $\dfrac{lb}{ft^3}$ in the FPS system. Mass density is measured in: $\dfrac{kg}{m^3}$, $\dfrac{g}{cm^3}$, and $\dfrac{slugs}{ft^3}$ in the MKS, CGS, and FPS systems of measurement respectively. A couple of practice problems will help you to tie the idea of density to your world of daily life. You get more practice with these ideas later in this chapter.

Newton's Figs

Working with a model can be confusing if you are not careful. Using the particle model, as is done here, means that if matter is portrayed as being made up of particles as described in the kinetic theory of matter, then certain things can be explained in terms of that model. If it is a good model, certain things can be predicted with a lot of success.

Physics Phun

1. A quantity of aluminum is 21 cm wide, 1.05 m high, and 4.03 m long. How much does the quantity of aluminum weigh if the mass density of aluminum is $2.70\,\dfrac{kg}{m^3}$? Hint: Remember that $\rho_w = \dfrac{w}{v} = \dfrac{mg}{v} = g\rho_m$.

Some solids like steel springs display the property of elasticity. Elasticity in a steel spring is the ability to stretch or distort within reason and then return to the original length. Elasticity depends on the cohesive forces of the particles of the solid. If the spring is stretched beyond its elastic limit, it will remain distorted and never returns to its original shape. You know that this property of elasticity is the same idea that you worked with on larger objects like the billiard balls.

You also observed the stretching of a spring when we developed the spring potential energy. Hooke's law is the basic idea for the elasticity discussed here as well. The force required to stretch the spring is directly proportional to the elongation of the spring as long as you remain within the elastic limit of the spring. You can hang a spring vertically and calibrate it to measure weight. Suppose that you have a spring supported vertically and hang 4.9 N of weight on the spring and observe an elongation (or stretch from the original length) of 0.14 m.

Using Hooke's law, what would you expect the elongation to be for 9.8 N? You can make this statement using Hooke's law: $\dfrac{4.9nt}{0.14m} = \dfrac{9.8nt}{y}$ where y is the unknown elongation. Do you recognize that statement as a *proportion?* Now solve the proportion for y. $(4.9N)y = 9.8N(0.14m)$, the $y = \dfrac{9.8N}{4.9N}(0.14m) = 0.28m$. What is the elongation for 14.7 N? What is the spring constant in the equation $F = ky$? Did you find that the spring constant is $35\dfrac{N}{m}$ and the elongation is $0.42m$?

Plain English

A **proportion** is an equation each of whose members is a ratio.

I hope you were able to do that. This is a little different than the application in which you were calculating spring potential energy but both deal with elasticity and Hooke's law. Both phenomena depend on those little bouncing balls or vibrating balls that are part of the particle nature of matter model of a solid.

Liquids

Like a solid, the particles of liquids vibrate rapidly, but in a random way. Unlike the vibration of the particles of a solid, which are about a point of equilibrium as if the solid particles are tied to that point by little elastic springs, the particles of the liquid vibrate in all directions haphazardly. The particles of a liquid are about as close to each other as those of a solid. However, they have more freedom of movement around and over each other with enough mobility to flow and take on the shape of their container. Liquids diffuse just as solids do, but whereas just a few particles of solids move from one solid to another, practically all of the particles of both liquids will diffuse.

You can demonstrate the diffusion of liquids by placing two liquids in a container with only the surfaces of the liquids in contact. That surface separates the liquids.

Placing a more dense colored liquid such as a blue copper sulfate solution in the container first and then the less dense liquid such as water on top of the dense substance can accomplish that. Your teacher can demonstrate this to you if you are in a class.

One of the liquids needs to be colored so that you can observe the diffusion that will be nearly complete in a few days. The more dense liquid diffuses upward and the less dense downward even in the presence of the force of gravity.

The particles of liquids cause cohesive forces as well as adhesive forces much like particles of a solid, except you will find that the forces are not as strong in liquids. As you observed with the stretch of a spring, a considerable force is required to stretch the spring and even more to exceed the elastic limit of the spring or to break it. You can observe the strength of the cohesive forces of a liquid by poking your finger in a thick liquid such as paint or motor oil. When you remove your finger, you experience a noticeable tug of the paint particles on your finger (the attraction of paint particles for paint particles) and the paint continues to adhere to your finger even after wiping (the attraction of paint particles for your finger).

You can see the relative strengths of the cohesive and adhesive force of water by pouring water into a graduated cylinder. The water particles tend to attract the glass particles more than they attract other water particles and cause a concave or crescent-shaped water surface. The water particles creep an observable distance up the sides of the cylinder, causing that part of the water surface to be higher than the middle of the column of water. If you observe a column of mercury in a glass tube, you find a convex surface on the top of the column because mercury particles attract each other more than they attract glass particles. The middle of the mercury surface is noticeably higher than the sides touching the glass.

The cohesive forces of particles in a liquid cause a liquid to have a free surface that is characterized by *surface tension*. At some time in your experience you have more than likely floated a needle on the surface of water. If you have not done the experiment, try this. Place a drinking glass filled with water on a level surface. Use an eye dropper to fill the glass until the surface of the water is actually above the glass rim. When the surface is perfectly still, gently slide a needle lengthwise onto the surface of the water in the glass. If you do not prick the surface with either end of the needle, it floats! Well, the water surface supports the needle so that it stays on the surface of the water. Add a drop of liquid soap very gently to the other side of the water. As it diffuses across the surface, it breaks apart the forces that cause the surface tension. Watch what happens to your needle. You can explain surface tension by thinking of the cohesive forces of the particles in the liquid acting on one particle. You know that the particles must be within a certain distance from each other to exert the cohesive force. If they are too close the force is repulsive and if they are too far away the force is negligible.

The particle of interest can be thought of as being affected by forces within a certain region surrounding the particle; call it the sphere of force. For a particle completely surrounded by particles of the liquid, the particles in its sphere of force will cancel each other out so the particle of interest has no net force. A particle of interest near to the surface will have its sphere of force decreased on top so there will be a slight unbalanced force downward because more particles attract it toward the liquid than those attracting it upward. A particle of interest at the surface, however, experiences only particles in its sphere of force pulling it toward the liquid. All particles at the surface experience this same pull, causing the surface to form a condition that resembles a membrane stretched taut over the free surface. Because of the tension, the surface tends to be as small as possible.

When the needle is placed on the inflexible filmlike surface, it makes a dimple in the surface, increasing the area. The cohesive forces tend to restore the minimum area, a taut horizontal surface, by exerting an equal but opposite force upward. As you found earlier in this chapter, liquids have the property of density the same as solids. The needle is denser than water, but it floats because of the surface tension. The water pushes up on the needle with as much force as the earth pulls down on it. Small bugs use this phenomenon to walk on lake water when it is still. Surface tension also causes small amounts of water such as dew or rain to form small spheres. The spherical surface is found to be the minimum surface for a given volume. The particles of the liquid attract the particles at the surface causing the surface to have the minimum surface possible with the appearance of an inflexible film stretched over the droplet.

Plain English

Surface tension is a quantity or condition of the surface of a liquid that causes it to tend to contract. **Pressure** is a quantity determined by the force on a unit of area.

Liquids share another property with solids: They exert *pressure*. Pressure is not force, it is force per area. It is a scalar quantity with units that you may have heard, like pounds per square inch (psi).

Pressure is exerted by a liquid in all different directions. You learn much more about pressure in the next chapter, but I want to discuss it briefly now before making a very detailed study of it later. You probably know a lot of people who use the words force and pressure interchangeably and now you can show them some distinct differences in the two concepts. To begin with, force is a vector quantity and pressure is not.

Next, look at the definition of pressure in symbols, $p = \dfrac{F}{A}$, and determine the units of measurement of this new quantity. In the FPS system, pressure is measured in $\dfrac{lb}{in^2}$ sometimes referred to as psi. The units of measurement in the MKS system are $\dfrac{N}{m^2}$ and in the CGS system $\dfrac{dynes}{cm^2}$. So the units of force and pressure are conceptually different, and as you discover other differences, you are justified in insisting on the distinction between pressure and force.

An example of the type of problem you can encounter involving pressure might be helpful here, especially if it emphasizes the difference in force and pressure.

Example 1: High Heel Pressure

Suppose that a 110 lb woman walks into the room wearing, among other things, high heel shoes. The heels have an area of about $\frac{1}{16} in^2$ at the base. Her full weight is first on one heel and then the other as she walks across the floor. How much pressure does she exert on the floor?

Solution:

$p = \dfrac{F}{A} = \dfrac{110lb}{\frac{1}{16} in^2} = 1760lb / in^2$! I left three significant figures in case the lady solves this problem. The correct answer is $1800lb/in^2$. That means that in this case a force of 110 lb causes a pressure of $1800lb/in^2$. It does emphasize a difference in magnitude of force and pressure. There is also a difference in units of measurement as well as the fact that force is a vector quantity and pressure is a scalar quantity.

Note: A key to understanding pressure is the area on which the force is pushing. A man in his street shoes exerts a much lower pressure on the road surface than the same man standing on ice skates exerts on the ice surface.

Example 2: Swimming Pool Pressure

Suppose a swimming pool is in the shape of a rectangular solid 4.00 m wide, 8.00 m long, and 2.00 m deep. It is completely filled with water. What is the pressure of the water on the bottom of the pool? Hint: The mass density of water is $10^3 \dfrac{kg}{m^3}$.

Solution:

$$p = \frac{F}{A_b} = \frac{W_w}{A_b} = \frac{\rho_{ww} V}{A_b} = \rho_{mw} g \frac{V}{A_b} = \frac{\left(10^3 kg / m^3\right)\left(9.8 m / s^2\right)(4.00m)(2.00m)(8.00m)}{(8.00m)(4.00m)}$$

$$= 1.96 \times 10^4 N / m^2$$

You notice that the pressure of a liquid depends only on the depth of the liquid.

That turns out to be about $409lb/in^2$. That is quite a bit less than the pressure by the lady. Do you know why? Try some of these problems on your own.

Physics Phun

1. A tank is 9.0 m long and 5.0 m wide. The tank is filled with milk to a depth of 4.5 m. The mass density of milk is **1.03g/cm³**. How much pressure does the milk exert on the bottom of the tank?

2. A water tank is 10.0 ft in diameter and 20.0 ft tall. What is the pressure on the bottom of the tank when it is completely filled with water? Hint: The volume of a cylinder is $V = \pi r^2 h$ and the mass density of water is **1g/cm³**. Hint: 3.28 ft = 100 cm.

3. A man weighs 192 lb. When he stands still supporting his weight on both shoes, the total area of his shoes in contact with the floor is **484cm²**. What pressure does he exert on the floor?

Gases

If you think the particles of solids and liquids are weird, get a load of these guys. The particles of a gas are so far apart that they do not exert force on each other until they bump into each other. Obviously, the particles are very far apart compared with those of solids and liquids. In fact, particles of matter in the gaseous state occupy about 10^3 times the volume that it occupies in the liquid state. That does not mean that the particles are bigger now that they are gaseous, just that they take up more space. It would be like a college student who gets very little living space in her dorm room. When she buys her first house, she thinks that she has much more room to move. She is taking up more space. Don't dare tell her that she has more volume now.

At room temperature, gas particles travel at about 500 m/s and travel about 100 nanometers before bumping into another particle or the sides of the container such as a sealed balloon or glass beaker. The sides of the container are experiencing between 4 and 10 billion collisions each second. Gases have density, exert pressure, and diffuse. Oh, do they diffuse! If someone walks into the room wearing wild perfume, you smell it almost immediately. The gas particles move randomly in all directions and mix with the other gas molecules in a container, the room, in nothing flat. These little critters move so fast and are so far apart that you can think of them as individual particles. It should be obvious by now that a gas does not have a definite volume or a definite shape. The particles move so fast in a random fashion that they completely fill their container.

A Johnnie's Alert

Even people who should know better confuse the words force and pressure. Remind yourself, with notes if necessary, that force is a vector quantity and pressure is a scalar quantity. Be able to identify the units of both in all systems of measurement.

The density of a gas is the same quantity as the density of a solid or liquid. Air has mass and weight. The mass density of air near sea level is found to be $1.29 \times 10^{-3} g / cm^3$. We are fortunate to live on a planet that surrounds us with a sea of gases, the atmosphere. The atmosphere that I am referring to is at or near sea level but the atmosphere of the earth extends upward for about 25 miles although most of it is below an altitude of 20 miles above the surface of the earth. Above 25 miles the atmosphere gets pretty thin. The pressure of the air we breathe near the surface of the earth is often reported as the pressure of a column of air one square inch in area and reaching to the top of the atmosphere.

The pressure at sea level of our atmosphere is 1 atmosphere, or 1 atm. That is the same as $14.7 lb / .in^2 = 1.013 \times 10^6 dynes / cm^2$. There are other measures of pressure, and I list some of them here for your information: $1\ Pa = 1$ Pascal $= 1 \dfrac{N}{m^2}$, $1\ bar = 1.000 \times 10^5 N / m^2 = 10^5 Pa$, $1\ atm = 1.013 \times 10^6 dynes / cm^2 = 1.013 \times 10^5 N / m^2$ $= 1.013 \times 10^5 Pa = 101.3 kPa = 1.013 bar$.

The unit of measure of pressure used depends on the application. Physicists will use the units reviewed earlier. The chemist will use the same units as the physicist along with millimeters of mercury, or Torr. The weatherman uses millimeters of mercury, inches of mercury, and bars or millibars when he discusses pressure in his weather reports. It is important for you to be able to interpret the pressure in a unit that is familiar. The pressure in millimeters of mercury or inches of mercury will be meaningful after a more complete discussion of pressure in the next chapter.

A Johnnie's Alert

Unless stated otherwise in a problem, the pressure near the earth's surface means the pressure at sea level at standard temperature $0°C$.

Changing State

At this stage, you have examined three different states of matter. You are very familiar with all three states of water, for instance, and know that they exist under certain conditions. In order to get ice to change state, some drastic changes must take place to alter the activity of the particles of the ice. That is, the particles of ice must be caused to vibrate so wildly that they start to move one over the other to become water. Another possibility is that, under the right conditions, the particles of ice become gas particles! The special name for that change is *sublimation*. You have probably observed the results of that process when you see what appears to be steam rising from the frost on a neighbor's roof in the early sunshine of a wintry morning. In reality what you see is condensed water vapor forming from the cold water molecules rising from the frost. You can't really see individual water molecules, you know.

Plain English _____

Sublimation is the direct change of a solid to a vapor without going through the liquid state. **Evaporation** is the production of a vapor or a gas from a liquid as some of the fastest-moving molecules escape from the liquid. Evaporation happens at any temperature. **Regelation** is the melting under pressure and then freezing again after the pressure is released.

A similar process is observed when the sun pops out after a summer rain shower. Steam appears to rise from the asphalt on rain-slick city streets. That occurs when the particles of a liquid suddenly become particles of a vapor. The process is called _evaporation_. While driving through the mountains in Colorado, I once saw clouds of vapor rising from the asphalt highway, caused by the sunshine on water from recently melted snow. The discussions of states of matter in this chapter make you aware that something drastic must be happening in order for the particles to make these drastic changes in water.

The ice skates of the figure skaters glide across the ice with such ease that it appears the skates are lubricated. A closer look at the tracks of the skates reveals that they are riding on a thin layer of water. You would notice a similar action if you observe a piece of bare wire with heavy weights on each end hanging across a block of ice. The action is not as fast as a gliding ice skate, but in time the wire cuts completely through the ice. The water immediately freezes above the wire, though, so that the block appears to remain intact even after the wire has passed through it. Again we note that in order to bring about such changes in matter, the particles must be experiencing sudden modifications of some sort to make matter exhibit that behavior. The name for this phenomenon is _regelation_. You discover some of the causes of these changes in the next two chapters. Think change of state and come up with your own explanation for the possibility of changing from a solid to liquid or liquid to a gas.

Boyle's Law

You may wonder why Boyle's law is a topic discussed now when there are several gas laws. Two reasons for my choice are (1) to emphasize the limitations of a mathematical model borrowed from mathematics by the scientist, and (2) to explore the relationship of two properties that we have discussed for all states or phases of matter: volume and pressure. You recall that the particles in solids and liquids attract each other at certain distances and if the distances are less the forces become repulsive. That means that solids and liquids cannot be compressed easily. The particles in a gas are so far apart that gases have no fixed shape or volume. The particles of a gas can be made to occupy a smaller volume with relative ease. In the larger scheme, gases differ from solids and liquids in that gases can be compressed much easier than liquids and solids.

The particles of solids differ from liquids and gases in that the particles of solids vibrate randomly about a point of equilibrium. The particles of solids do not have the freedom of motion that the particles of liquids and gases do. Because of the freedom of motion of the particles of liquids and gases, liquids and gases are referred to collectively as fluids. The streamlined flow of fluids was mentioned when we looked at the difference in pressure on an airplane wing as an application of Bernoulli's principle. That picture should become clearer now that you have a better understanding of pressure. That picture will gain more clarity as we add a few more tiles to the mosaic in the next chapter.

A known mass of gas in a closed container and constant temperature will behave according to the mathematical model shown by the graph in Figure 13.1. Using the graphical analysis of data outlined in an earlier chapter in this book, you can see that Figure 13.1 suggests an inverse variation between the volume of a gas and the pressure that it exerts if the temperature and amount of gas particles remain constant.

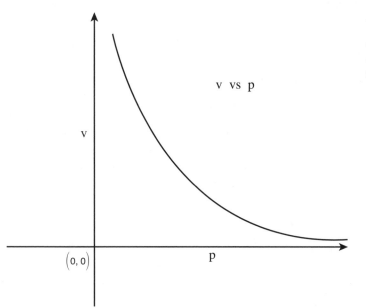

Figure 13.1

A graph of volume vs. pressure reveals a mathematical model called Boyle's law.

In fact, as the data is presented in that graph, you can see that apparently $V \propto \dfrac{1}{p}$. You know that you can see just how closely the data fits that mathematical model by plotting V vs. $\dfrac{1}{p}$. A straight-line graph reveals that you have a good model to fit the data, and in addition the slope of the graph is the constant of proportionality. It will be no surprise to you, after your work with this book, to find that Boyle's law is not the absolute truth. That is, it is not always true. You know that if the pressure is increased enough so that the particles start affecting each other by attraction the gas starts behaving more like a liquid and less like a gas. Boyle's law breaks down at that point because it is a model for a gas.

When you see a mathematical statement of a scientific law or principle, you know that the statement does not mean the law or principle is true now, always has been true, and always will be true. The scientist identifies models that best fit the observations of his or her experimentations with the constraints of his or her laboratory procedures. If you keep the limitations of science in mind, you will enjoy the adventure of exploring scientific thought. You may even be able to identify limitations and create ways to explain phenomena in a way to go beyond those shortcomings. However, you will not be disappointed to find even new explanations will have limitations. The excitement of this adventure continues to unfold.

The Least You Need to Know

- ◆ Identify at least three properties of a solid, a liquid, and a gas.
- ◆ Calculate the pressure of a liquid at a fixed depth.
- ◆ Identify the units of pressure in each system of measurement.
- ◆ List three ways that force differs from pressure.
- ◆ Describe how particles can be used as a model to explain solids, liquids, and gases.

Pressure

In This Chapter

- ◆ Pressure and high heels
- ◆ The deeper the water the greater the pressure
- ◆ The higher you go the thinner the air
- ◆ Pressure and baseball
- ◆ The art of compressing air

Many times I have watched tugboats push barges loaded with wheat along the channel of the huge Colombia River. The wheat is piled so high that the barges seem to barely float. They do float and enough of the barge is below the surface of the water to float the load safely down the river. Ships at sea obey the same laws of physics to haul cargo from one continent to another. It is fascinating to realize that the barge or ship must sink far enough into the water to support the craft and cargo.

How much an object must sink into a liquid to float is an interesting problem. In this chapter, you will explore the calculations of the amount of sinking in order to float. You will also find out how much a balloon can lift when it is inflated with a gas. These problems involve the notion of pressure, which I promised that I would discuss in more detail in this chapter.

Is Pressure a Force?

By now, you know the answer to this question. Pressure is not a force. It is closely related to force, as you will find in this chapter. First of all pressure is defined to be force per unit area. Then you calculated the force on the bottom of a pool due to the weight of the water and from that you calculated the pressure on the bottom. So you are familiar with the relationship between force and pressure, and know that they are not equivalent.

Pressure and Liquids

Pressure is exerted in all directions in a fluid. Suppose you have a confined fluid, oil maybe, in a container with a cylinder at each end both completely filled with the fluid like that found in Figure 14.1.

Figure 14.1

The hydraulic press operates as a simple machine.

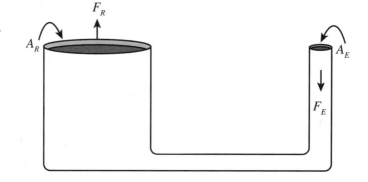

A Johnnie's Alert

Because pressure is transmitted undiminished throughout a liquid, liquid exposed to the atmosphere will have the pressure of the atmosphere transmitted undiminished throughout the liquid. That fact is taken into account when calculating total pressure in a liquid. Many problems involving pressure in liquids specify only the pressure of the liquid.

On top of each cylinder is a piston that is free to move up and down. This arrangement constitutes a simple machine that will multiply force by using pressure on the confined motionless fluid. The machine is based on Pascal's principle which states that any pressure applied to a confined fluid at rest will be transmitted undiminished to every point in the liquid. That means that the pressure will cause a force to act perpendicular to every unit of surface area exposed to the confined liquid.

So if you push down on A_E, the pressure transmitted undiminished throughout the fluid will cause a force upward on A_R. This arrangement acts like a simple machine and has applications like the lift that raises your car when the mechanic changes the oil or switches the tires. Compare this machine to other simple machines.

Notice that when the effort piston is pushed down a distance s_E, a volume of the fluid is displaced. Since the pressure is transmitted undiminished throughout the liquid, an equal volume of liquid is displaced in the resistance cylinder. That means that the resistance piston moves a distance s_R to displace an equivalent volume. Since the volume in each case is the volume of a tiny cylinder of fluid, multiplying the area times the distance moved defines the volume of the fluid displaced in both the effort cylinder and the resistance cylinder.

CAUTION

A Johnnie's Alert

Your mechanic probably applies the effort force by releasing compressed air into a chamber that causes the effort piston to move. You never see the details discussed here in play at the service station, but you can see the giant cylinder move upward, raising your car on the service ramp.

$$V_E = V_R$$

$$A_E s_E = A_R s_R$$

$$\frac{s_E}{s_R} = \frac{A_R}{A_E}$$

$$IMA = \frac{A_R}{A_E} \quad \text{Definition of IMA, the ideal mechanical advantage.}$$

$$IMA = \frac{(d_E)^2}{(d_R)^2} = \frac{(r_E)^2}{(r_R)^2} \quad \text{Since the pistons are circular and the area of the circular}$$

piston is proportional to the square of its diameter and the square of its radius.

$$p_E = p_R$$

$$\frac{F_E}{A_E} = \frac{F_R}{A_R} \quad \text{Definition of pressure.}$$

$$\frac{A_R}{A_E} = \frac{F_R}{F_E}$$

You see that pressure can play an important role in the operation of a simple machine to multiply force even though pressure is not a force. Oh, you do not see that it multiplies force? Consider the following example problem.

Example 1: Pressure and a Simple Machine

Suppose your mechanic lifts your two-ton SUV (a ton is the name of 2000 pounds of weight) with his hydraulic lift. The area of the large piston is $1130 in^2$ and the area of the small piston is $13 in^2$. How much effort force must be applied to the small piston?

Solution:

$$F_R = 4 \times 10^3 \, lb$$

$$F_E = ?$$

$$A_R = 1130 \, in^2$$

$$A_E = 13 \, in^2$$

$$\frac{A_R}{A_E} = \frac{F_R}{F_E}$$

$$F_E = \frac{A_E}{A_R} F_R$$

$$F_E = \frac{13}{1130} \left(4 \times 10^3 \, lb \right)$$

$$F_E = 46 \, lb$$

So to lift a two-ton SUV requires a force of only 46 lbs. Remember, in order to multiply force with a simple machine, distance must be sacrificed. That means the liquid in the small diameter pipe moves a much greater distance than in the large diameter pipe. Also, the areas of the pipes are directly proportional to the square of their diameters or radii. Pressure is not a force, but it is closely related to force as you see here and in the rest of this chapter.

Liquid Pressure and Depth

Do you think that a dam with a lake 10 miles long behind it must be stronger than a dam with a lake 5 miles long behind it? That sounds like another stupid question, doesn't it? Well, if you do not have an answer now, you will have by the time you finish this section.

Newton's Figs

Pressure acts in all directions at a point in a liquid. The pressure in a liquid is the same at every point in a horizontal plane at a given depth in the liquid.

You have noticed that when you dive into a swimming pool and stand on the bottom, your ears experience a great deal of pressure. It does not matter how you turn your head, you experience the same pressure. That is consistent with the earlier discussion of pressure: Pressure is exerted equally in all directions at a given depth. You probably noticed that you had to travel pretty deep in the swimming pool to experience much pressure. Does the pressure depend on how deep you are in the water? Refer to Figure 14.2. I use that diagram to develop a way of calculating the pressure at any depth of a liquid.

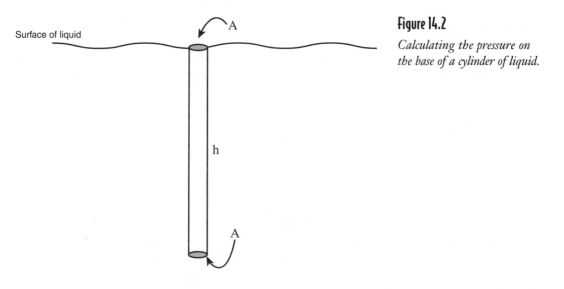

Surface of liquid

A

h

A

Figure 14.2

Calculating the pressure on the base of a cylinder of liquid.

The top of the cylinder is at the surface of the liquid. The bottom of the cylinder is at a depth h in the liquid. The cylinder is just a cylinder of the liquid itself. You know from the last chapter that the pressure on the bottom of the cylinder is the weight of the liquid divided by the area of the base of the cylinder. The cylinder is uniform, with the area of the base the same as the area of the top of the cylinder. You calculate the pressure on the bottom in the following way:

$$p = \frac{F}{A}$$

$$p = \frac{W_L}{A}$$

$$p = \frac{\rho_L V_L}{A}$$

$$p = \frac{\rho_L A h}{A}$$

$$p = \rho_L h$$

Newton's Figs

Checking the units of measurement implied by an algebraic solution or resulting from an arithmetic solution can often reveal errors in reasoning and/or omissions of factors or terms of sums. For example, if your solution should be dynes and you find that grams are the actual units, you probably omitted a factor of the acceleration due to gravity, $980 cm/s^2$.

The final result indicates that you can calculate the pressure of any liquid at any depth for which the weight density is constant. The pressure at any depth can be calculated by multiplying the weight density of the liquid by the depth.

Pressure and Total Force

Remember when you calculated the total force on the bottom of the swimming pool as the weight of the water? Now you can calculate the total force on the bottom of the pool as:

$$p = \frac{F}{A}$$

$$F = pA$$

$$F = \rho_w h A$$

$$F = \rho_w V$$

Newton's Figs

The average value of a quantity that is directly proportional to another quantity can be calculated by adding its smallest value to its largest value, then dividing the sum by two.

What do you know about that? That is the same answer you found before. It is comforting to know that you can find the same answer by using a new result you just derived. But wait, there's more. What do you suppose the total force would be on the sides? As long as the swimming pool is in the shape of a rectangular solid, you can handle it. If the shape is an ellipse or if the pool varies in depth, you will probably need calculus to find the total force on the sides.

However, the procedure that you will use on a swimming pool with a simple shape is the same procedure you need to use for a weird shape. Since the pressure varies with the depth, you must use the average pressure to find the total force on a side. That is the part of the whole truth that I could not tell you when you first calculated the force on the bottom of the pool. You can now savor the fruits of delayed gratification.

$$p_{avg} = \frac{F_{total}}{A_{side}}$$

$$F_{total} = p_{avg} A_{side}$$

$$F_{total} = \rho_w h_{avg} A_{side}$$

You should check the units in that last equation and apply that idea to the solution of a problem in order to get a clearer picture of force as related to pressure.

Example 2: Total Force

Suppose you have a swimming pool full of water. The swimming pool is 20.0 ft long, 10.0 ft wide, and 6.00 ft deep. Calculate the total force on the bottom of the pool and the total force on the ends and the sides due to the water in the pool.

Solution:

$$\rho_{wL} = 62.4 lb / ft^3$$

$$l = 20.0 ft$$

$$w = 10.0 ft$$

$$h = 6.00 ft$$

$$F_{total\ end} = ?$$

$$F_{total\ sides} = ?$$

$$F_{total\ bottom} = ?$$

A Johnnie's Alert

The pressure of a liquid varies directly as the depth, and the constant of proportionality is the weight density of the liquid. The weight density of a liquid varies in some liquids, such as water in the ocean.

$$F_{total\ bottom} = \rho_{ww} h A_{bottom}$$ By the results of your work in this section.

$$F_{total\ bottom} = \rho_{ww} hwl$$

$$F_{total\ bottom} = \left(62.4 lb / ft^3\right)\left(6.00\,ft\right)\left(10.0\,ft \times 20.0\,ft\right) = 7.49 \times 10^4\,lb$$

$$F_{total\ end} = P_{avg} A_{end}$$

$$F_{total\ end} = \rho_{ww} h_{avg} A_{end}$$

$$F_{total\ end} = \rho_{ww} \frac{h}{2} wh$$

$$F_{total\ end} = \left(62.4 lb / ft^3\right)\left(10.0\,ft\right)\frac{\left(6.00\,ft\right)^2}{2} = 1.12 \times 10^4\,lb$$

$$F_{total\ side} = \rho_{ww} h_{avg} A_{side}$$

$$F_{total\ side} = \rho_{ww} \frac{h}{2} lh$$

$$F_{total\ side} = \left(62.4 lb / ft^3\right)\frac{\left(6.00\,ft\right)^2}{2}\left(20.0\,ft\right) = 2.25 \times 10^4\,lb$$

There, you have painted in more details to the picture of the role that pressure plays in this type of situation.

There may be a couple of things bothering you about this solution. The average depth is calculated by adding the shortest depth to the largest depth and dividing by two, that is, $h_{avg} = \dfrac{0 + h}{2} = \dfrac{h}{2}$. The force on the ends and sides starts out as zero at the minimum depth and increases as the depth increases to the greatest value at maximum depth.

Since the pressure of the liquid is exerted in all directions, you can think of the force due to the pressure as being perpendicular to the ends and sides all the way from top to bottom. There is more of this story.

Pressure and Buoyant Force

Refer to Figure 14.3 and think of the small uniform cylinder as completely submerged in a liquid.

Figure 14.3

The cylinder completely submerged in water has a buoyant force.

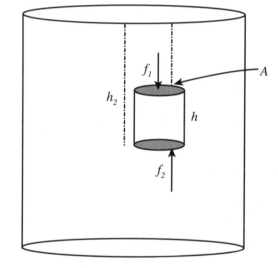

Use water for the liquid. You know that, due to the pressure of the liquid, there is a force, f_2, acting upward on the bottom of the submerged cylinder that is at a depth h_2. For the same reason, there is a force, f_1, with direction downward on the top of the submerged cylinder. Supposedly Archimedes yelled, "Eureka!" and then ran bare down the street after discovering that there is a net force on a submerged object because of the pressure differences in the liquid. You are about to find whether Archimedes was justified in declaring that naked truth. Since pressure varies directly as the depth of a liquid, you know that $f_2 > f_1$, so there must be a net force due to the liquid.

$$f_2 - f_1 = f_{net}$$
$$\rho_{ww} h_2 - \rho_{ww} h_1 = f_{net}$$
$$\rho_{ww}(h_2 - h_1) = f_{net}$$
$$\rho_{ww} h = f_{net} \quad \text{Because } h_2 - h_1 = h, \text{ the height of the submerged cylinder.}$$
$$W_{liquid\ displaced} = f_{buoyant}$$

Eureka! The weight of the liquid displaced by the submerged cylinder, or any object for that matter, is equal to the buoyant force on the submerged object due to the pressure differences in the liquid.

Suppose the object is not completely submerged. You know that the object must float in the liquid and is not moving vertically. That means that there is no net force on the floating object. Refer to Figure 14.4 to see a cylindrical object that has length, h, and uniform cross section, A_{cyl}, floating in a liquid with h_a out of the liquid and h_L submerged in the liquid.

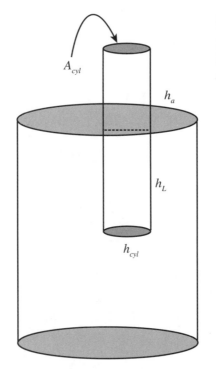

A_{cyl}

h_a

h_L

h_{cyl}

Figure 14.4

The floating cylinder in a liquid that provides a buoyant force.

The buoyant force on the object must have the same magnitude as the weight of the object. That is, if the liquid is water,

$$f_b = W_{cyl}$$

$$\rho_{ww} V_{w\,disp} = \rho_{w\,cyl} V_{cyl}$$

$$\rho_{ww} A_{cyl} h_L = \rho_{w\,cyl} A_{cyl} h$$

$$\rho_{ww} h_L = \rho_{w\,cyl} h$$

$$\frac{h_L}{h_{cyl}} = \frac{\rho_{w\,cyl}}{\rho_{ww}}$$

The LST is made of steel and so are submarines and other ships, and yet they float very well. It does not matter what an object is made of as long as it displaces enough water so that the buoyant force of the water is greater than the weight of the object floating in the water.

That means that as long as the ratio of the amount of the object submerged to the total amount of the object is less than one, the object will float. That will happen as long as the weight density of the object is less than the weight density of the liquid, water in this case. Examples of some floating objects are the oil-laden ships in the Persian Gulf and the work-horse LSTs that ferried our soldiers along with their equipment to shore during World War II.

Notice that if the liquid is water, the last equation, $\frac{h_L}{h_{cyl}} = \frac{\rho_{w\,cyl}}{\rho_{ww}}$, enables you to determine the *specific gravity* of the object.

Plain English

The **specific gravity** of a substance is the ratio of the weight density of the liquid to the weight density of water. It is also the ratio of the mass density of the liquid to the mass density of water. The specific gravity of a liquid is a ratio using water as a standard. The specific gravity of a gas is a ratio using dry air at S.T.P. as a standard. The **hydrometer** is an instrument used to measure the specific gravity of a liquid.

That is, $\frac{h_L}{h_{cyl}} = \frac{\rho_{w\,cyl}}{\rho_{ww}} = spgr_{cyl}$. You now have all the information you need to construct an instrument that will measure the specific gravity of a liquid with a uniform object like the cylinder in Figure 14.4, given that it floats in water. You can use symbols that enable you to generalize the discussion a bit such as h_{sL} to represent the amount submerged in a liquid and h_{sw} to represent the amount the same object is submerged in water. The expression then becomes, $\frac{h_{sL}}{h_{cyl}} = \frac{\rho_{w\,cyl}}{\rho_{wL}}$. Then:

$\rho_{w\,ccyl} = \frac{h_{sL}}{h_{cyl}} \rho_{wL}$ Multiplying both members by ρ_{wL}.

$\frac{\rho_{w\,cyl}}{\rho_{ww}} = \left(\frac{h_{sL}}{h_{cyl}}\right)\left(\frac{\rho_{wL}}{\rho_{ww}}\right)$ Dividing both members by ρ_{ww}.

$spgr_{cyl} = \frac{h_{sL}}{h_{cyl}} spgr_L$ Definition of specific gravity.

$spgr_{cyl} = \frac{h_{sL}}{h_{cyl}}$ If the liquid is water.

$$\dfrac{spgr_{cyl} = \dfrac{h_{sL}}{h_{cyl}}\,spgr_{L}}{spgr_{cyl} = \dfrac{h_{s\,water}}{h_{cyl}}} \quad \text{Dividing both members by } spgr_{cyl}.$$

$$1 = \dfrac{h_{sL}}{h_{s\,water}}\,spgr_{L} \quad \text{Simplifying.}$$

$$spgr_{L} = \dfrac{h_{s\,water}}{h_{sL}}\,! \quad \text{Multiplying both members by } \dfrac{h_{s\,water}}{h_{sL}}.$$

Therefore, we may use a uniform rod, glass tube, floating cylinder, etc., to determine the specific gravity of any liquid by dividing the length submerged when placed in water by the length submerged when placed in any other liquid. The device you have just designed is called a *hydrometer*. The hydrometer has several useful applications, from the determination of the condition of your car battery to the determination of the correct content of alcohol in the brewing of wine or beer.

Physics Phun

1. Suppose that the surface of the lake behind Grand Coulee dam is 445 ft above the base of the dam. What is the pressure of the water at the base of the dam? Would a lake 10 miles long behind the dam have twice the pressure as one 5 miles long? What is the approximate total force on one square inch of the dam at the base? Now you know why that dam is 500 ft wide at the base and 30 ft wide at the top.

2. A bundle of crushed tin cans weighs 10.5 lb in air and 9.10 lb in water. Calculate the specific gravity of tin.

3. A rectangular flat river barge is floating empty in a river. The barge is 25 ft wide and 105 ft long. How much deeper will it sink into the water when 255 tons of wheat are loaded on it?

Gas Pressure and Altitude

You know that a gas is a fluid but not the same as a liquid. They are both fluids in that they are free to flow and take on the shapes of their containers. A liquid has a definite volume but a gas expands and completely fills its container and has neither a definite shape nor volume. A liquid exerts pressure, and a gas can exert pressure.

A Johnnie's Alert

Gases, like liquids, exert pressure in all directions at a point. The pressure of a gas is the same at all points on a horizontal plane at a given level. That means that everything, even the human body, near the surface of the earth is experiencing 14.7lb/in^2 of pressure on every square inch of the body.

You already know that a gas exerts pressure but in a different way than a liquid. The particles of the gas collide with the sides of the container many times each second. In each collision, the particles and sides exchange impulses. The pressure of a gas is the result of the billions of impulses of the bouncing particles. You also know that you can increase the pressure of a gas by placing more particles into the container. You will find another way of increasing the pressure of a gas in the next chapter. You already know how to measure the pressure of a liquid and since a gas is a fluid, maybe you can use that knowledge to find a way to measure the pressure of a gas.

Sipping Cider Through a Straw

The atmosphere (atm) is a good source of gas, or a mixture of gases, with which to begin. You found earlier that at sea level the pressure of the atmosphere is 14.7lb/in^2. Do you know how that was measured? We will find a way to do that but begin with an activity with which you are familiar.

You have sipped cider through a straw at some time in your life. You do that by removing the air from the straw, and the cider rises through the straw to tickle your taste buds. If you think about it, the liquid does not rise up through the straw but is pushed up by the atmospheric pressure on the surface of the liquid outside the straw. When you removed the air inside the straw with your mouth and cheeks, there is little or no opposition to the pressure of the atmosphere so the liquid is pushed up the straw to fill the region of reduced pressure. That is an idea for measuring the pressure of the atmosphere; measure the pressure of the atmosphere by relating it somehow to the pressure of a liquid.

Torricelli did just that. He invented the mercurial barometer more than 300 years ago. For many years, I demonstrated his invention to my classes. Handling mercury is hazardous to your health and can be hazardous to the health of those in the same room, so don't try this at home. Now they tell me. A glass tube with a hole at one end is filled with mercury from a small reservoir of mercury; a microscope slide is placed over the open end and held firmly in place. The tube is turned upside down in the reservoir until the open end with the microscope slide cover is below the surface of the mercury in the reservoir. When it is certain that the open end is under the surface of the mercury, the microscope slide is slowly removed while the tube is held above the bottom of the reservoir but below the surface of mercury. As you expect, the mercury flows out of the tube into the reservoir until the weight of the mercury in the tube is just the right amount to balance the upward force due to atmospheric pressure on the surface of the reservoir of mercury and transmitted through the liquid mercury. The space above the mercury offers essentially no resistance because it is now nearly a perfect vacuum just like the straw you use to sip cider. The mercury had occupied the empty part of the tube before it ran out into the reservoir.

The height of mercury in the tube stands at 760 mm, 76.0 cm of Hg or 29.92 in of Hg when the measurement is made at sea level and when the temperature is 0°C. When you work with gases, the conditions S.T.P., standard temperature and pressure, are 76.0 cm of mercury and 0°C.

When 1 atm was defined as 14.7*lb/in²*, I said that I would explain what is meant by 76.0 cm of mercury, 760 mm of mercury, or about 30 inches of mercury. Now you know that the pressure at sea level can be expressed in terms of the height of a mercury column that is balanced by 1 atm of pressure. The 760 mm of mercury is a direct measure of atmospheric pressure even though it is not directly in *lb/in²* or *dynes/cm²*.

The mercurial barometer is diagrammed in Figure 14.5, showing that 1 atm of pressure on the reservoir of mercury causes the mercury to rise to a height of 76.0 cm inside the glass tube.

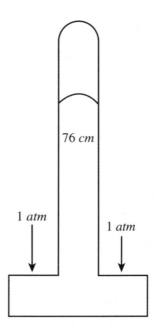

76 cm

1 atm

1 atm

Figure 14.5

The mercurial barometer measures the atmospheric pressure directly.

Pascal demonstrated the change in pressure with altitude by having a mercurial barometer taken to the top of a high mountain. He found that the height of the column of mercury was significantly less at the greater altitude.

A Johnnie's Alert

The greater the distance from sea level, or at increased altitude, the smaller the reading on a barometer. The lower pressures at increased altitudes results from a lower density of air particles and the fact the weight of the column of air above you is now less so the pressure is less. The nearly zero density beyond our atmosphere in space is very nearly a vacuum.

There are fewer particles of air per unit volume at increased elevations so there is less pressure on the surface of mercury in the mercurial barometer. You may wonder why mercury is used and not a more common liquid like water. As it turns out, mercury has a specific gravity of 13.6. You know that means the weight density of mercury is 13.6 times the weight density of water. That means mercury is a liquid that is readily available and has the largest density of common liquids. If you use water, you find that the height of water would be about 34 ft for standard pressure, or 13.6 times higher than the column of mercury. Thirty inches times 13.6 is 408 inches. That is 34 ft. A height of 34 ft is not very practical unless you have a giant available to read it.

It is worthwhile for you to make a few of the calculations suggested by the facts shared with you so far in order to clarify your understanding of the quantity that is referred to as the pressure of the atmosphere.

$$p = \rho_{ww}h_w = \left(62.4 lb \,/\, ft^3\right)\left(34 ft\right) = \left(2122 lb \,/\, ft^2\right)\left(\frac{1\, ft^2}{144\, in^2}\right) = 14.7\, lb \,/\, in^2$$

$$p = \rho_{wm}h_m = \rho_{ww}spgr_m h_m = \left(980\, dynes \,/\, cm^3\right)\left(13.6\right)\left(76.0 cm\right) = 1.013 \times 10^6\, dynes \,/\, cm^2$$

Remember that 1 atm = $14.7\, lb \,/\, in^2$ = $1.013 \times 10^6\, dynes \,/\, cm^2$ = 1.013 bars. You know that not only does a gas exert pressure, but you can also measure atmospheric pressure directly with a barometer.

Air Pressure and Buoyant Force

The atmosphere exerts pressure, and it also has density. The density of a gas is measured at S.T.P. (76.0 cm of mercury and 0°C), and at S.T.P. the mass density of dry air is $1.29 \times 10^{-3}\, \frac{g}{cm^3}$. You find that there are times when the density of a gas is $1.29 \frac{g}{l}$; read 1.29 grams per liter. That means that you need to be familiar with the liter as a measure of volume.

I list some defining equations here for you to use in comparing familiar measures of volume with this new measure.

$1\, ml = 1\, cm^3$ Read one milliliter equals one cubic centimeter.

$10^3\, ml = 10^3\, cm^3$ Multiplying each member by 10^3.

$1\, l = 10^3\, cm^3$ One liter is equal to one thousand milliliters.

You can use these defining equations to interpret the new unit in terms of a familiar unit of measurement as follows:

$$\left(1.29\frac{g}{l}\right)\left(\frac{1\,l}{10^3\,cm^3}\right)=1.29\frac{g}{10^3\,cm^3}=1.29\times10^{-3}\frac{g}{cm^3}$$ Notice that a defining equation was used to define unity in order to change the looks of $\frac{g}{l}$ to $\frac{g}{cm^3}$.

Since a gas has density, it behaves like other fluids in that objects submerged in a gas experience a buoyant force. Consider a safe gas to handle, helium, and use the buoyant force of air to calculate how much a liter of He, the symbol for helium, will raise or lift. You need the specific gravity of helium or the weight density of helium. Usually it is easy to look up the specific gravity of a solid, a liquid, or a gas. It may be difficult to find the weight density of those substances.

The specific gravity of a gas is defined to be the ratio of the weight density of a gas to the weight density of air at S.T.P.

The weight density of helium can be expressed in terms of the specific gravity of helium that is given to be 0.138. Specifically, $\rho_{wHe}=spgr_{He}\rho_{wA}$, the weight density of helium is equal to the product of the specific gravity of helium and the weight density of air.

A Johnnie's Alert

Properties of gases are often given in terms of S.T.P., 760 mm of mercury and 0°C, and volume is often given in liters. I have given you information in this part of the text that enables you to interpret these quantities in terms of more familiar quantities discussed earlier.

Newton's Figs

Buoyant force is calculated for gases the same way it is in liquids. Since they are both fluids, you can say the buoyant force is the weight of the fluid displaced by the object immersed in the fluid.

$F_B-W_{He}=F_L$ In words, the equation states that the buoyant force minus the weight of the helium, actually helium and container, is equal to the lifting force.

$$\rho_{wA}V_{He}-\rho_{He}V_{He}=F_L$$

$$\rho_{wA}V_{He}-\rho_{wA}spgr_{He}V_{He}=F_L$$

$$V_{He}\rho_{wA}\left(1-spgr_{He}\right)=F_L$$

$$(1\,l)\left(1.29\frac{g}{l}\times980\frac{cm}{sec^2}\right)(1-0.138)=F_L$$

$$F_L=1.09\times10^3\,dynes$$

That means that one liter of helium will lift 1090 dynes of weight at S.T.P. How many newtons is that? How many pounds would that be? You can probably handle such questions but just in case I list a couple of mass and weight relations and a defining equation here and make some conversions to help you out:

454 grams weigh one pound on earth.

1 kilogram weighs 2.2 pounds on earth.

Therefore, 9.80 N = 2.2 lb, or one pound equals 4.45 N.

$$1\,dyne = 1\frac{g-cm}{s^2} = \left(\frac{1\,g-cm}{s^2}\right)\left(10^{-3}\frac{kg}{g}\right)\left(10^{-2}\frac{m}{cm}\right) = 10^{-5}\frac{kg-m}{s^2} = 10^{-5}\,N$$

$$1\,dyne = 10^{-5}\,N = \left(10^{-5}\,N\right)\left(\frac{2.2\,lb}{9.80\,N}\right) = 2.24\times10^{-6}\,lb$$

You can figure out answers to questions like that all by yourself from now on.
$$1.09\times10^{3}\,dynes = 1.09\times10^{3}\times10^{-5}\,N = 1.09\times10^{-2}\,N = 2.45\times10^{-3}\,lb.$$

Does a Baseball Curve?

Remember when we talked about the lift of an airplane wing? The idea is that particles of fluids move at different speeds around or through objects suspended in the fluid, immersed in the fluid, or containing a fluid. Bernoulli's principle helps you to understand the behavior of such objects subjected to a streamline flow of fluid. Remember the principle states $\frac{KE}{V} + p =$ constant. That means that particles traveling at a higher speed will have a larger KE, so the first term of the equation is larger; then the second term, p, must be smaller. Consider the flight of the baseball in Figure 14.6, which shows the ball traveling toward the left of the diagram, indicated with the large arrow.

Figure 14.6

The baseball in flight tends to move toward the bottom of the diagram.

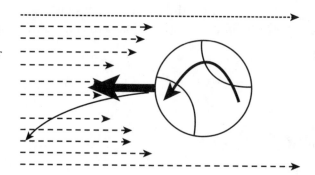

The ball experiences a breeze blowing toward the right, shown as dotted lines with arrowheads pointing to the right of the diagram. If you swing your arm through the air, you can feel the breeze the ball is experiencing. The ball is spinning counterclockwise as indicated in the diagram. Particles of air near the baseball are rotating counterclockwise with the baseball as the seams drag the air particles along with the spinning ball. The particles near the top of the baseball, that part of the ball near the top of the page, are traveling in a direction opposite the breeze and are slowed down by the opposing motion of the breeze. The particles near the bottom of the baseball are traveling in the same direction as the breeze and are sped up by those particles. That means that the pressure at the bottom is less than the pressure at the top of the baseball, and so the baseball is pushed toward the region of lower pressure. The baseball travels in a curved path as indicated down and to the left in the diagram. The diagram is oversimplified but shows how a baseball curves. The direction of the curve will depend on your frame of reference.

Newton's Figs

Think of Bernoulli's principle as being composed of two terms: the first term increases with the speed of particles in a streamline flow, and the second term, pressure, is added to the first term to yield a constant value. If the first term increases because of higher speeds of particles, the pressure must decrease in order for the sum to be constant.

If the flight of the ball is viewed from above while the ball spins in a horizontal plane, the baseball would curve away from a right-handed batter. If viewed from the side while the ball rotates in a vertical plane, the baseball would curve downward as it approaches home plate. The curved path of the ball is exaggerated in the diagram to make the point but the ball does curve enough that you can see batters duck and dodge it as if they are being attacked by a drunken hornet!

Pressure and Thermal Energy

Thermal energy is discussed in more detail in the next chapter but your general education has provided you with knowledge of thermal energy. If you have ever removed a steamy apple pie from an oven and accidentally spilled some of its savory juice on a naked hand, you know what thermal energy is! What does that have to do with pressure? You have probably experienced a flat tire on a car at some time or other. By the way, you know that you do not measure the pressure of the air in a tire with a mercurial barometer. The gadget used is called an air gauge. If you have a flat tire in some lonely spot and must inflate the tire with an air pump, you not only get a fantastic workout, but you also cause the pump to increase in temperature in the region where the air is forced into the tire. After removing the nail that caused the flat, and inserting enough air in the tire to get you on your way, you probably will have to do the same thing over several times until you reach a service station. The air pump gets hotter each time you use it.

You may not even use a manual air pump for that purpose. Maybe pumping up a football or basketball is all you have ever used a manual pump for. Anyway, an air pump increases in temperature for a good reason—one that you have probably already guessed. Since a gas is compressible, the pump takes in air at atmospheric pressure and compresses it, decreasing its volume significantly.

In order to decrease the volume, the particles of the gas must be forced to be closer together. You force the particles to be closer together by doing work on them with the pump. Much of the work done appears in the form of thermal energy. Sure there is some thermal energy as a result of the friction caused by the operation of the air pump, but most of the thermal energy is generated by the work done to compress the air that is forced into the leaking flat tire. Your curiosity about the new idea of thermal energy will be better satisfied in the next chapter. Thanks for playing the flat tire game with me for the sake of illustration. Of course, I know that you are the type who just takes the flat off and replaces it with a spare tire. I hope your spare is not flat.

Newton's Figs

Thermal energy is the result of doing work upon a gas when it is compressed. The more work done compressing the gas, the more thermal energy is developed.

The Least You Need to Know

- ◆ Explain what is meant by standard temperature and pressure.
- ◆ Calculate the pressure of a liquid at any depth.
- ◆ Find the buoyant force of fluids.
- ◆ Interpret 76.0 cm of mercury as a measure of pressure.
- ◆ Apply the density of fluids in problem solving.

Chapter 15

Heat Energy

In This Chapter

- ◆ Measuring temperature
- ◆ Degrees Celsius or Celsius degrees
- ◆ Quantities of thermal energy
- ◆ Thermal energy required to change ice to water
- ◆ How heat gets from here to there

You already know when something is hot or cold. The thermal energy of a body is the kinetic and potential energy of the particles of the body. A hot body can transfer heat to a cold body and when it does, its thermal energy is reduced. The thermal energy of the body absorbing the heat is increased.

In this chapter, you will use temperature to identify a body that is hot or cold and become familiar with three different scales for measuring temperature. Keep in mind that thermal energy is conserved in the transfer of heat from an object with a higher temperature to an object having a lower temperature.

What Is the Temperature Today?

Remember when I discussed taking the apple pie from the oven and experiencing pain when some of the liquid landed on your arm? I mentioned that as knowledge most of us gained at some time in our lives. Everyone has an idea about what heat is; we typically know when something is hot. Those are two entirely different ideas that I discuss with you in this chapter.

You know that when you boil a pan of water the water gets hot. You may not have realized that if you pour some of the hot water in a cup and set the pan on a hot pad near the cup, the cup of hot water does not have as much *heat* as the several cups of hot water left in the pan.

You know that the chemical reaction that takes place during the burning of a match feels hot, but it will not increase the temperature of the room like a rolling, crackling fire in the fireplace on a cold, damp Portland evening. You also know that there is nothing more refreshing than the cooling ice cubes in a glass of lemonade while enjoying the summer sun on a Chesapeake beach. The ice cubes do not have as high a temperature as the lemonade so the ice cools the drink by absorbing heat from the lemonade.

We began our examination of solids, liquids, and gases by considering the particles of each as if they acted like little elastic balls bouncing around randomly. In solids, they are more like elastic balls attached to elastic springs that keep pulling them back toward a point of equilibrium. The particles in liquids and gases have more freedom of movement but they are in constant motion in all states of matter. The total of the kinetic energy and potential energy of those little particles as they bounce and vibrate is the energy we call internal energy or thermal energy.

Plain English _____

Heat is the energy a body at a higher temperature transfers to a body of lower temperature by reducing the thermal energy of the body that had the higher temperature initially. The thermal energy of a body is the kinetic and potential energy of the particles of which the body is composed. **Temperature** is the condition of a body that tells how hot the body is compared to other bodies. Bodies with high temperature are in a condition to transfer thermal energy to bodies with lower temperatures. Bodies with low temperature are in a condition to absorb thermal energy.

If you increase the agitation of those particles about which we were talking, you increase their total internal energies so the body containing those particles has more thermal energy. You can increase the motion of the particles by causing a chemical reaction like striking a match placing them so that they are exposed to a radiant source like the sun, bending a piece of wire back and forth several times, or passing an electrical current through wire like the filament of a light bulb. You can decrease the energy of the particles by bringing them close to an object with a lower temperature like the ice cubes in the lemonade.

You know that you gave the pie heat by placing it in the oven. As the pie takes on more heat, the *temperature* of the pie increases. The ice cubes in the lemonade absorb thermal energy from the drink giving it that refreshing sensation on a hot day. The juice from the pie had a very high temperature and that is why the tiny drop that hit your hand gave some of its thermal energy to your lower-temperature body. An ice cube placed on the hot spot on your skin gives instant relief by absorbing some of the thermal energy from the little blister. The ice cube is in a condition to take on or absorb thermal energy and the recently scalded skin is in a condition to release thermal energy readily. Accidents like this can be avoided by a little thought about temperature and what thermal energy is. Hot pans of water and hot pies must be guarded carefully when little children are around because you do not want them to learn about thermal energy by an accident. That is why one of the first words they learn is "hot."

Thermometers

Temperature can be measured with the proper instrument but thermal energy is more difficult to determine quantitatively. Actually we really never know the whole story of how much thermal energy is in something. We only know how much is transferred. Consider the measurement of temperature first and then we will spend some time calculating the quantity, Q, the symbol for thermal energy. You probably are familiar with the thermometer as the instrument for measuring temperature. You may not know that there are four different scales for temperature. We discuss three of those scales because mostly engineers use the other scale.

Most people in the world are interested only in the absolute or Kelvin scale and the centigrade or Celsius scale. I include the Fahrenheit scale because that is the scale used widely for the measurement of temperature in the United States. I also choose to use ideas that help you over some of the rough spots in physics and avoid terminology that may be confusing. My reason is that I am sharing ideas to help you understand physics and not to challenge you to be a rocket scientist. That may be your next step after completing this book. Therefore, you find that I list readings on the Celsius scale as 20°C, on the Fahrenheit 20°F, and on the Kelvin scale 20 K. (Notice that no degree mark is used with the Kelvin scale.) Please do not think that these three temperatures are the same. You would be comfortable at 20°C, freezing at 20°F, and dead stiff at 20 K.

You are well aware of the fact that people in the United States recognize 68°F as a comfortable room temperature while people in other countries recognize 20°C as a comfortable room temperature. I will feel remiss if I do not discuss both. Scientists, and the rest of the world, use the Celsius scale and the Kelvin scale for their calculations and news weather reports.

Effects of Change in Temperature

When heat is added to most materials, they expand. Glass and mercury expand when heated but mercury expands more than glass. The thermometer is calibrated to read temperature as the difference in the expansion of mercury and glass.

> **CAUTION**
>
> ### A Johnnie's Alert
>
> Water has many properties that make it a unique substance. When liquid water releases heat, it becomes denser until it reaches about 4°C, and then gets less dense as it continues to release heat. As water changes to ice it continues to release thermal energy. The ice expands with such a tremendous force that it can break the blocks of cars that are not properly protected by antifreeze. The expansion is so drastic that ice floats in water even though most of the ice is below the surface. This is not typical of most solids in contact with their liquids. Most materials are denser in their solid state than in their liquid state.

The thing that you measure in a thermometer is the height of the mercury column inside the small-diameter sealed glass tube. Water is the standard for calibrating the thermometer because it has a definite point at which it freezes, called the *freezing point of water*, and at which it boils, at standard pressure, called the *boiling point of water*. The procedure used to calibrate a thermometer is to place the thermometer in water and let the water freeze. Then etch the place where you see the top of the mercury on the glass. Next place the thermometer in water and let it boil, at standard pressure. Finally, etch the place where you see the top of the mercury on the glass.

Of course, today we use alcohol in many thermometers rather than mercury to avoid the problems that come from broken thermometers and escaping mercury.

Comparing Temperature Scales

The freezing point on the Celsius scale is named 0°C, on the Fahrenheit scale it is designated 32°F, and on the Kelvin scale it is named 273 K. The boiling point on the Celsius scale is labeled 100°C, on the Fahrenheit scale it is labeled 212°C, and on the Kelvin scale it is designated 373 K.

Temperature Change or Change in Temperature?

The remainder of these scales is discussed later, but at this time you should be aware of two different references that you make to the temperature scale. One is the reading that you record by seeing which mark best lines up with the top of the column of mercury in the glass tube. You can observe that the thermometer reads $0°C$ when the bulb of the thermometer is in freezing water and that it reads $100°C$ when placed in boiling water at standard pressure. You observed a temperature change from a reading of $0°C$ to a reading of $100°C$. The other reference is that you observed a change in temperature of $\Delta t = 100°C - 0°C = 100C°$. Notice the difference in reporting these two observations; the first observation is two different readings and the second observation is 100 Celsius degrees change in temperature. You make calculations later that will require you to recognize the distinction between $°C$ and $C°$.

The first is a temperature change, one or two or more readings, and the second is a change in temperature, the difference in two readings. Having these two things clearly in mind helps you to avoid confusion when you use them in calculations. Getting back to the markings on the temperature scales: The Celsius scale and the Kelvin scale have equal marks on the thermometer for the smaller divisions of degrees. There are $100°C$ between the freezing point and boiling point on the Celsius scale and 100 K between the freezing point and boiling point on the Kelvin scale. The Fahrenheit scale is a little weird in the same way as the FPS system. There are $180°F$ between the freezing point and the boiling point on the Fahrenheit scale. That creates a little problem for converting from one scale to the other, but it is solved later.

First look at the conversion from Celsius readings to Kelvin readings. You convert $20°C$ to a Kelvin reading by adding 20 K to 273 K and get a reading of 293 K. The symbol that I will use for Kelvin temperatures in the following equations is T. That means that I can summarize the conversion as $20°C = T_{fp} + \Delta T = 273K + 20K = 293K$. Similarly, 300K is converted to a Celsius reading by finding the change in temperature for the freezing point to 300 K, which is 27 K, and adding that change in temperature to the freezing point of the Celsius scale to find $27°C$ is the correct Celsius reading.

 A Johnnie's Alert _____

Converting a Celsius reading to a Kelvin reading involves the calculation of a change in temperature from the freezing point to the reading to be converted. The change in temperature is in $C°$ The $C°$ is equal to the K degree so adding the change in temperature from the freezing point to the reading on the Celsius scale to the freezing point on the Kelvin scale results in the correct Kelvin reading. The same procedure is followed when converting from Kelvin to Celsius readings.

Converting readings between Celsius and Fahrenheit requires some thought. The approach to a solution to this problem is much like the conversion between Celsius and Kelvin scales. Calculate the ratio of the change in temperature from the freezing point to a reading on each scale and then set that equal to the ratio of the change in temperature between freezing point and boiling point on each scale. In symbols, that is: $\frac{t_C - 0°C}{t_F - 32°F} = \frac{100°C - 0°C}{212°F - 32°F}$. Reading the proportion you see that it says in words, a change in temperature on the Celsius scale is to the corresponding change in temperature on the Fahrenheit scale as a known change in temperature on the Celsius scale is to the corresponding change in temperature on the Fahrenheit scale. It is important that you see that there are no tricks and no magic, just a straightforward statement about two different scales. Notice that both ratios in the proportion have the units $\frac{C°}{F°}$ and if we agree that t_C is a Celsius reading and t_F is a Fahrenheit reading, then we can write: $\frac{t_C}{t_F - 32} = \frac{100}{180}$, $t_C = \frac{5}{9}(t_F - 32)$, or $t_F = \frac{9}{5}t_C + 32$ If you want to convert Fahrenheit readings to Celsius readings use the first equation. You can convert Celsius readings to Fahrenheit readings with the second equation. You can check by using a reading that is the same on both scales, $-40°F = -40°C$.

Check the first equation $t_C = \frac{5}{9}(-40 - 32) = \frac{5}{9}(-72) = -40$. That means that $t_C = -40°C$ because we agreed that t_C must be a Celsius reading in order to simplify the equations.

Check the second equation $t_F = \frac{9}{5}(-40) + 32 = -72 + 32 = -40$ so $t_F = -40°F$ by our agreement that t_C is a Fahrenheit reading. Some people like to do the arithmetic in their heads to make these conversions and some people remember these two equations to accomplish that. Another thing that you can do is take the original proportion and simplify it so that you only remember one equation.

That is: $\frac{t_C}{t_F - 32} = \frac{100}{180}$, $t_C = \frac{5}{9}(t_F - 32)$, $9t_C = 5t_F - 160$,

and $9t_C - 5t_F + 160 = 0$.

A Johnnie's Alert

Check a conversion equation or any algebraic equation that you intend to apply a lot by using values that you understand. I chose −40 because I know that the reading on the Fahrenheit scale and the Celsius scale have the same magnitude at that temperature.

You should check this equation also. Convert −40°F to a reading on the Celsius scale. You may simplify the last equation a bit before checking. You can write it as $9C - 5F + 160 = 0$. Use the last equation to check: $9C - 5(-40) + 160 = 9C + 360 = 0$, $9C = 360$, and $C = -40$. It works! You may use any method you choose to convert readings but make sure you can convert readings from Fahrenheit to Celsius and from Celsius to Fahrenheit.

Any one of these thermometers may be used to measure temperature so long as the reading is above $-39°C$ because mercury freezes at that temperature. At very high temperatures glass melts so even the thermometer has its limitations for measuring temperature.

Newton's Figs

One way of thinking about converting readings between Celsius and Fahrenheit scales is to realize that $100C° = 180F°$ and $5C° = 9F°$. That means to convert $25°C$ to a Fahrenheit reading realize that the Celsius reading is $25C°$ from the freezing point, and that is the same as $\left(25\left(\dfrac{9}{5}\right)\right)F°$ or $45F°$ from the freezing point or a reading of $(45 + 32)°F = 77°F$.

Where the Celsius and Fahrenheit scales use the freezing point of water and the boiling point of water for calibration, the Kelvin scale starts at the lowest temperature possible, absolute zero. Measurements only as low as about $-273.16°C$ have been observed. As you may expect, for extreme temperatures special devices must be used. A thermocouple is an electrical device that serves very well in the measurement of extreme temperatures. You are not expected to be able to deal with a thermocouple, but you should know that there are means of dealing with the limitations of the glass-mercury thermometer.

Specific Heat

You know that temperature is the condition of a body to transfer thermal energy or to absorb thermal energy. Also, thermal energy is the energy transferred from one body to another because of a difference in temperature. You are ready to find how much heat a body has. The measure of the temperature of a body helps to find a quantity of thermal energy. The body has other properties that must be taken into account when you determine the heat of a body.

Returning to the model of matter briefly, you know that the particles of matter are vibrating or bouncing around randomly at rather high speeds. When objects are warmed, the particles move even more rapidly and tend to cover more distance. That is observed as expansion of the body or object. When thermal energy is removed from a body, the body is observed to contract. The amount of expansion and contraction is not very great and for solids and liquids the expansion per unit length or volume depends upon the type of matter. Gases are observed to expand and contract but the amount of expansion per unit volume is about the same for all gases.

Water is a substance that behaves abnormally as far as expansion and contraction are concerned. When thermal energy is added to ice, it changes to water at 0°C. As more thermal energy is added to the ice water, the particles move closer together and are closest at 4°C. Water has it greatest density, $1.00 g / cm^3$, at 4°C because the mass remains constant. As the water absorbs more thermal energy, the particles move farther apart. One implication of this strange behavior is that the densest water in a lake or pond sinks to the bottom. The less dense water is at or near the top at a temperature of 0°C, where it freezes when enough thermal energy is removed from it.

If it were not for this strange property of water, you could see water freezing from the bottom of the lake or pond to the top. No doubt ice-skating would be out and ice-fishing would be a sight to behold! When you say that warming a body causes the particles to move more rapidly, keep in mind that may mean that such motion involves changing from a solid, ice, to a liquid, causing the strange behavior of water.

Before you examine other properties of a substance that must be considered in order to measure thermal energy, look at some new units of measurement that are involved. Since thermal energy is energy that is transferred because of a difference in temperature, it can be measured in ergs, joules, or ft-lb but many writers use a different unit applied especially to heat. Notice that strange substance, water, is the standard for the new units. You find in many reference books that physicists use the SI, MKS, or CGS units, but I will list other units that may be used.

First, the *calorie* (cal) is the quantity of heat required to raise the temperature of one gram of water one Celsius degree. This unit of heat is sometimes referred to as the small calorie to emphasize the difference between the calorie and the Calorie, or so-called large calorie. The MKS unit of heat is the *kilocalorie* (kcal), which is also referred to as the Calorie, the unit of heat used by dieticians and some biological scientists to measure the fuel value of foods. As is usually the case for the FPS system, the unit of heat is strange; it is the Btu. The *British thermal unit* (Btu) is the quantity of heat required to raise the temperature of one pound of water one Fahrenheit degree.

Plain English _____

The **calorie** is the quantity of heat needed to raise the temperature of one gram of water one Celsius degree. The **kilocalorie** is the quantity of heat required to raise the temperature of one kilogram of water one Celsius degree. The **British thermal unit** is the quantity of heat required to raise the temperature of one pound of water one Fahrenheit degree. The **heat capacity** of a body is the quantity of heat required to raise its temperature one degree. The **specific heat** of a substance is the ratio of its heat capacity to its mass or weight.

Make special note that all three units are defined in terms of a change in temperature. The metric units are defined in terms of mass and temperature changes in the Celsius or Kelvin scales, and the FPS unit is defined in terms of weight and temperature changes in the Fahrenheit scale. Since the size of the degree in the Celsius scale is the same as the degree in the Kelvin scale, it does not matter which scale you use when you are talking about a difference in temperatures. The size of the difference between zero degrees Celsius and 100°C is the same as between 273 K and 373 K. In the last section of this chapter, I outline the relationship between work and heat that was developed by James Prescott Joule. From that relationship, you find that 1 cal = 4.19 joule and 1 kcal = 4190 joules. It is helpful for you to know a relationship between units of thermal energy in the FPS system and the metric system, as well as SI units, 1 Btu = 252 cal = 1056 joules. The contractors who discuss the needs of your home for air conditioners and furnaces usually talk about the number of Btu's involved. That is another reason why I discuss the measurement of heat in different units.

Another property of an object that contributes to the measurement of heat is called its *heat capacity*. You have observed that when adding heat to some objects, it takes longer to change the temperature of some materials than to change the temperature of others.

Those objects that take longer to increase their temperature also take longer to cool off than others do. This property of a substance is called its heat capacity. The heat capacity of a substance is the heat necessary to raise the temperature of a material one degree.

The heat capacity of an object can be expressed as: heat capacity $= \dfrac{Q}{\Delta t}$, with units of $\dfrac{kcal}{C^\circ}, \dfrac{cal}{C^\circ}$, or $\dfrac{Btu}{F^\circ}$.

Heat capacity does not depend on the type of material or the mass of the substance.

A more useful property of a substance that takes into account the type of substance and mass is the *specific heat* of an object.

The specific heat depends on the particular type of matter as well as the mass of the object. Specific heat is usually represented as $c = \dfrac{Q}{m\Delta t}$ with units $\dfrac{kcal}{kgC^\circ}, \dfrac{cal}{gC^\circ}$, or $\dfrac{Btu}{lbF^\circ}$. Finally you have one way of calculating a quantity of thermal energy. Suppose you have a 5.00 g piece of aluminum, $c = 0.214\dfrac{cal}{gC^\circ}$, that is warmed up from 5.00°C to 55.0°C. How much heat is required to cause the change in temperature? The specific heat gives you a start: $c_{al} = \dfrac{Q}{m_{al}\Delta t}$, $Q = c_{al}m_{al}\left(t_f - t_0\right)$, and $Q = \left(0.214\dfrac{cal}{gC^\circ}\right)(5.00g)\left(55.0^\circ - 5.00^\circ C\right)$

$= 53.5cal.$

The problem you just completed had the change in temperature as part of the information given. You know that thermal energy leaves objects of higher temperature and enters objects of lower temperature until their temperatures are the same. In fact, that process is formalized in the Second Law of Thermodynamics as pointed out before.

Plain English

The **First Law of Thermodynamics** states that the heat absorbed by cold substances is equal to the heat released by hot substances if there is no thermal exchange with the environment and no work is done on or by the system.

There is also a First Law of Thermodynamics, which states the thermal energy absorbed by cold substances is equal to the thermal energy released by the hot substances if there is no thermal exchange with the environment and no work is done on or by the system. One law tells you about the amounts of thermal energy that are exchanged, and the other tells you which way the exchange takes place. This law applies whenever hot and cold substances are mixed or brought into contact. A symbolic statement of the law is $Q_{gained} = Q_{released}$.

Any time you have hot substances or objects and you bring them into contact, you know that no heat energy is lost from the mixture if they are thermally isolated from their environment. Anyone knows that if you leave that glass of lemonade with ice in it on the porch overnight, by the morning the ice will have melted and the lemonade will be at the temperature of the outside air. The thermal energy will have been given up to the outside air if it is not isolated from its environment. In an isolated system, any thermal energy that is lost by hot objects or substances is gained by cold objects or substances. An application of this law helps you to see how this works.

Example 1:

During an experiment conducted in a laboratory, a student mixed 231 g of aluminum at 92.0°C with 64.2 g of water at 24.2°C in a Styrofoam cup with a cap. What was the final temperature of the mixture?

Solution:

$m_{al} = 231\,g$

$t_{al} = 92.0°C$

$m_w = 64.2\,g$

$t_w = 24.2°C$

$c_{al} = 0.214$

$c_w = 1.00\,\dfrac{cal}{gC°}$

$t_f = ?$

$Q_{gained} = Q_{released}$, $m_w c_w (t_f - t_w) = m_{al} c_{al} (t_{al} - t_f)$, because you know $t_f > t_w$ and $t_{al} > t_f$,

$(64.2g)\left(1.00 \dfrac{cal}{gC°}\right)(t_f - 24.2°C) = (231g)\left(0.214 \dfrac{cal}{gC°}\right)(92.0°C - t_f)$, $64.2t_f - 1554 =$

$4548 - 49.4t_f$, $64.2t_f + 49.4t_f = 4548 + 1554$, $114t_f = 6102$, and $t_f = 53.5°C$.

If you substitute this value for t_f you find that the answers for heat released and heat gained differ by about two percent. If you solve for t_f algebraically first, you find that the two quantities of heat differ by less than one percent. That is another good reason to solve algebraically first and then substitute arithmetic values to calculate one answer. Substituting arithmetic values early in the solution will introduce errors due to rounding numbers in intermediate calculations before reaching a final solution. Consider the following: $t_f = \dfrac{m_{al} c_{al} t_{al} + m_w c_w t_w}{m_w c_w + m_{al} c_{al}}$, as the complete algebraic solution, which demonstrates another advantage.

The numerator of the fraction has units $(g)\left(\dfrac{cal}{gC°}\right)(°C)$ and the denominator has units $(g)\left(\dfrac{cal}{gC°}\right)$ so the quotient yields $°C$ as the units of the final answer. Substituting arithmetic values into the algebraic solution gives you:

$t_f = \dfrac{(231)(0.214)(92.0) + (64.2)(1.00)(24.2)}{(64.2)(1.00) + (231)(0.214)} = 53.7°C$. If you substitute this answer for

t_f, you find that the heat gained differs from the heat released by less than one percent. Both answers are correct but the last answer is better because the rounding errors have been reduced considerably, making our final result more convincing that

$Q_{gained} = Q_{released}$. Solve a few problems of this type to gain some confidence in your ability to handle the new laws of heat exchange.

Physics Phun

1. Find the final temperature of the mixture if 21.0 g of water at 5.00°C are mixed with 42.0 g of water at 42.0°C.

2. Find the final temperature of the mixture if 2.00 lb of water at 41.0°F are mixed with 4.00 lb of water at 91.0°F.

3. A piece of iron weighing 4.00 lb and having a temperature of 209°F is dropped into 2.00lb of water having a temperature of 42°F. What is the specific heat of iron if the final temperature of the mixture is 71°F?

Latent Heat of Vaporization and Fusion

There are other steps of calculating quantities of heat besides the method involving the specific heat. One method of calculating a quantity of thermal energy involves a change of state. Thermal energy can be added to a solid and the temperature of the solid rises until it reaches a certain characteristic temperature. At that temperature, the solid changes to a liquid. The temperature at which this happens is called the *melting point*. The thermal energy continues to be absorbed, but the temperature does not increase. That must mean that the internal kinetic energy is not increasing. Where is the energy going then? Remember those little spring-like forces that were holding the solid together? Those attractions are being stretched. Work is being done. The internal potential energy of the material is increasing just like it did if you pulled on a spring and stretched it. The process of changing from a solid to a liquid is called *fusion*. Crystalline solids, like ice, have a definite melting point, and the *freezing point* is the same temperature. A liquid becomes a solid during a process called *solidification*.

Plain English

The **melting point** is the temperature at which a solid changes to a liquid. The process of changing from a solid to a liquid is called **fusion or melting**. Fusing in this case means the same as when it is used to mean melting metals together to make an alloy. The **freezing point** is the temperature at which a liquid changes to a solid. **Solidification, or freezing,** is the process of changing from a liquid to a solid.

A Johnnie's Alert

The melting point and freezing point is the same temperature for most crystalline substances. Noncrystalline substances do not have a definite melting or freezing point. Such substances freeze slowly and melt slowly at no specific temperature.

The temperature at which solidification takes place is called the freezing point. The melting point of ice is $0°C$ and the freezing point of water is $0°C$ but you know that even though the temperature is the same the processes are very different.

In order to change ice from your freezer to a liquid, in other words to melt ice, heat must be added to the ice until it is at $0°C$; then you must continue to add thermal energy to the ice until it changes to water at $0°C$. The amount of thermal energy you must add to ice at $0°C$ to change it to water at the same temperature is called the latent heat of fusion, or just the heat of fusion. Each crystalline solid has a characteristic heat of fusion. The heat of fusion for ice is $80 \dfrac{cal}{g}$ or $144 \dfrac{Btu}{lb}$.

That means that you must add 80 cal of thermal energy to every gram of ice at $0°C$ to change it to water at $0°C$ or 144 Btu to every lb of ice at $32°F$ to change it to water at $32°F$.

In like manner, water must be cooled until it reaches 0°C, and then 80 cal of thermal energy must be removed from each gram of water to change it to ice at 0°C. Notice that you use the specific heat of ice to find the quantity of heat necessary to warm it up to the melting point, then calculate the quantity of thermal energy necessary to change it to water by using the heat of fusion. Similarly, you use the specific heat of water to find the quantity of heat required to cool the water to the freezing point, and then determine the quantity of thermal energy necessary to change it to ice by using the heat of fusion. You know how to calculate quantities of thermal energy in two different steps, and a third step will be discussed after you have had some practice.

> **A Johnnie's Alert**
>
> The heat of fusion is the quantity of thermal energy required to change a solid to a liquid or a liquid to a solid without changing the temperature of the final substance.

Example 2:

Suppose you have 50.0 g of ice, specific heat 0.500 cal / g $C°$, at –20.0°C and you change it to water at 20°C. How much thermal energy is required for the job?

Solution:

$m_i = 50.0g$

$c_i = 0.500cal / gC°$

$t_i = -20.0°C$

$m_w = 50.0g$

$c_w = 1.00cal / gC°$

$t_w = 20.0°C$

$L_f = 80cal / g$

$Q = ?$

$$Q = m_i c_i \left(0°C - t_i\right) + m_i L_f + m_w c_w \left(t_w - 0°C\right)$$

In verbal summary: The quantity of thermal energy equals the amount of heat required to warm the ice to the melting point plus the amount of thermal energy required to change state plus the amount of heat required to warm the water from the melting point to the final temperature.

$$Q = (50.0g)(0.500cal / gC°)(20.0C°) + (50.0g)(80cal / g) + (50.0g)(1.00cal / gC°)(20.0C°)$$

$$Q = 500cal + 4000cal + 1000cal = 5500cal$$

The calculation of the required thermal energy in this type of problem is straightforward and the algebraic statement of the problem provides a good bookkeeping plan to make sure you do not leave anything out. I included the temperature at the melting point in the algebraic statement as a reminder that the ice becomes water at the melting point, and the water is warmed up from the melting point to the final temperature given in the problem. I am sure you welcome some practice on your own.

Physics Phun

1. Calculate the final temperature of a mixture of 8.00 g of ice at 0.00°C and 60.0 g of water at 90.0°C.
2. Calculate the quantity of thermal energy required to change 15.0 lb of ice at 20.0°F to water at 40.0°F.

You can calculate the quantity of heat needed to change the temperature of a material using specific heat if you want to warm a substance up or cool it off. You can also calculate the quantity of thermal energy necessary to change a solid to a liquid or a liquid to a solid. If you use specific heat, remember the quantity of heat depends on the substance, the mass of the substance, and the change in temperature. Calculating the quantity of thermal energy for changing from a solid to a liquid depends upon the substance, its heat of fusion, and its mass—there is not a change in temperature. One other quantity of thermal energy you can calculate involves changing state from a liquid to a gas or from a gas to a liquid. The new quantity of thermal energy depends upon the type of material and the amount of it you have. The change of state takes place at the boiling point.

A liquid must absorb heat until it reaches its boiling point. That requires the use of the specific heat of the particular liquid, of course. Then you must continue adding thermal energy to the liquid at the boiling point until the liquid changes to a gas or a vapor at the boiling point. The process is called *boiling*. Changing from a gas to a liquid is just the reverse process and is called *condensation*.

Plain English

Evaporation is the process of a liquid changing to a gas or a vapor. **Boiling** is the process of a liquid changing to a gas or a vapor at its boiling point. **Condensation** is the process of a vapor or a gas changing to a liquid.

The gas is cooled off until it reaches the boiling point. Calculating the quantity of heat removed from the gas to cool it to the boiling point requires the use of the specific heat of the gas. By continuing to remove thermal energy from the gas or vapor at the normal boiling point (the boiling point at standard pressure), you cause it to condense, and it becomes a liquid at the boiling point.

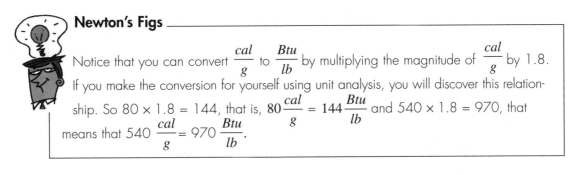

Newton's Figs

Notice that you can convert $\frac{cal}{g}$ to $\frac{Btu}{lb}$ by multiplying the magnitude of $\frac{cal}{g}$ by 1.8. If you make the conversion for yourself using unit analysis, you will discover this relationship. So $80 \times 1.8 = 144$, that is, $80\frac{cal}{g} = 144\frac{Btu}{lb}$ and $540 \times 1.8 = 970$, that means that $540\frac{cal}{g} = 970\frac{Btu}{lb}$.

The quantity of thermal energy required to change a liquid to a vapor or a gas is called the latent heat of vaporization, or *heat of vaporization*.

The heat of vaporization is the quantity of thermal energy required to change a unit mass or weight of a liquid to a vapor or gas at the normal boiling point. The heat of vaporization for water is 540 *cal/g* or 970 *Btu/lb*. That means that once the temperature of water is increased to the normal boiling point of 100°C, you must continue adding thermal energy to change the water to steam at the same temperature. The normal boiling point for water is 100°C or 212°F at one atmosphere of pressure. Did you realize that water will boil at a lower temperature when it is at a lower pressure, such as those that can occur at higher altitudes? That is why you will see special high-altitude cooking instructions on the recipes for some cake mixes. Water is used in the example, but any liquid may be used. However, that requires the use of a reference book to find the heat of vaporization for other liquids. Consider this problem:

Plain English

The **heat of vaporization** is the quantity of thermal energy required to change a unit mass or weight of a liquid to a gas or vapor at the normal boiling point.

Example 3:

Suppose you have 50.0 g of water at 50.0°C, and you want to change it to steam, specific heat 0.500 *cal/gC°*, at 110°C. How much thermal energy is required to complete the task?

Solution:

$m_w = 50.0g$

$c_w = 1.00 cal / gC°$

$t_w = 50.0°C$

$L_v = 540 cal / g$

$c_s = 0.500cal \, / \, gC°$

$t_s = 110°C$

$Q = ?$

$$Q = m_w c_w \left(100°C - t_w\right) + m_w L_v + m_s c_s \left(t_s - 100°C\right)$$

$$Q = (50.0g)\left(1.00cal \, / \, gC°\right)\left(100°C - 50.0°C\right) + (540cal \, / \, g)(50.0g) +$$
$$(50.0g)\left(0.500cal \, / \, gC°\right)\left(110°C - 100°C\right)$$

$$Q = 2500cal + 27000cal + 250cal = 29800cal$$

You can calculate a quantity of thermal energy in three steps. Specific heat is used to calculate the heat required to warm or cool a substance. The heat of fusion, L_f, enables you to find the quantity of thermal energy required to change a solid to a liquid or a liquid to a solid at the melting point. Finally, the heat of vaporization, L_v, is used to determine the quantity of thermal energy required to change a liquid to a gas or to change a vapor or a gas to a liquid at the normal boiling point. Remember that the temperature does not change when a change of state occurs.

It is a good idea to write an algebraic statement for the quantity of thermal energy to provide a plan to follow when you calculate the thermal energy involved. The algebraic statement can help you to see easily each quantity of thermal energy required. It also provides you with a method of accounting for each quantity of thermal energy so that every quantity of thermal energy required is included in your calculations. You know that the solids and liquids considered so far are in some kind of container. I have not included a container before, but I do in the next example so that you have a more realistic picture of handling some of the substances. Steam can be contained also; the procedure for working with steam in a container is the same as that used for containing ice and liquids.

CAUTION
A Johnnie's Alert

Many people lose terms in their calculation of a quantity of thermal energy by not relying on an algebraic statement of the problem. The algebraic statement is much like a recipe for your favorite dish—if you leave out an important ingredient, the whole thing is garbage.

Example 4:

You are given ice and its container both at −15.0°C. The mass of the ice is 55.5 g and the aluminum container, specific heat 0.214 *cal/gC°*, has a mass of 88.4 g. The ice remains in the container as it is changed to steam. How much thermal energy is required to change the ice to steam? Hint: The melting point of aluminum is almost 700°C, so the aluminum does not change state in this problem!

Solution:

$t_i = -15°C$

$t_c = t_i$

$m_i = 55.5g$

$c_i = 0.500cal / gC°$

$c_w = 1.00cal / gC°$

$m_{al} = 88.4g$

$c_{al} = 0.214cal / gC°$

$L_f = 80cal / g$

$L_v = 540cal / g$

$Q = m_i c_i(0°C - t_i) + m_{al}c_{al}(0°C - t_i) + m_i L_f + m_i c_w(100°C - 0°C) + m_{al}c_{al}(100°C - 0°C) + m_i L_v$

$Q = (55.5g)(0.500cal / gC°)(15C°) + (88.4g)(0.214cal / gC°) + (55.5g)(80cal / g) +$

$\quad + (55.5g)(1.00cal / gC°)(100C°) + (88.4g)(0.214cal / gC°)(100C°) + (55.5g)(540cal / g),$

$Q = 416.25cal + 283.764cal + 4440cal + 5550cal + 1891.76cal + 29970cal = 42600cal$

Physics Phun

1. Suppose you have 16.0 lb of steam at 212°F and it condenses, cools, and changes to ice at 32°F. How much thermal energy does this process release? Hint: $L_v = 970\dfrac{Btu}{lb}$, $L_f = 144\dfrac{Btu}{lb}$, $c_w = 1\dfrac{Btu}{lbF°}$.

2. You are given 46.0 g of ice at −25.0°C, held by a 75.0 g copper container, specific heat 0.0921 $cal/gC°$, and it is to be changed to steam at 100°C. How much thermal energy is required? Hint: $c_{ice} = 0.500\dfrac{cal}{gC°}$, $c_{water} = 1.00\dfrac{cal}{gC°}$, $L_v = 540\dfrac{cal}{g}$, copper remains a solid.

3. An aluminum container weighs 0.50 lb and holds 2.00 lb of water at 80.0°F. What is the final temperature of the mixture when .046 lb of steam at 212°F is added to the container of water?

Transfer of Heat

Three methods of thermal energy transfer are *conduction*, *convection*, and *radiation*. You can hear the weatherman discuss the weather in terms of these different methods of thermal energy transfer.

Plain English _____

Conduction is the transfer of thermal energy within a substance from one particle to the next while the particles are not moved from one place to another. Thermal energy causes the particles to vibrate more and the vibrating particles bump into neighboring particles, transferring the energy throughout the object. **Convection** is the transfer of thermal energy by the movement of matter. **Radiation** is the transfer of thermal energy by having only the energy transferred. No substance or convection currents are needed.

Transfer of Heat by Conduction

Have you ever held a needle in the flame of a match to purify it before removing a splinter from your finger? If you have, you probably have also experienced a painful finger that held the needle in the flame because it gets very hot. The thermal energy from the match is transferred to your finger through the needle. This type of transfer of thermal energy is called conduction. The vibrating particles in the substance transfer their energy to neighboring particles, sending the thermal energy throughout the object.

The needle mentioned earlier is a good conductor of thermal energy. Not all substances are good conductors. Substances that are not good conductors of thermal energy insulate you from the winter cold in your house and in your clothes.

Transfer of Heat by Convection

Thermal energy is also transferred by convection. You know that hot air rises because it is less dense and cold air rushes down to take its place. Cold water entering your hot water heater moves to the bottom of the tank where it is heated. The warm water in the tank rises and supplies warm or hot water to your home. The circulating motion of the water in the tank is an example of the transfer of heat by convection. The air that is warmed in your fireplace rises up the chimney, carrying away the smoke and much of the thermal energy in the warm air.

Transfer of Heat by Radiation

Some of the thermal energy from the fireplace is transferred to the room to warm your back when you stand in front of it. The fire does not warm you by conduction because that usually means contact with the source. There is warming by convection because of the rising hot air and falling cold air, but that is not what causes you to get too hot if you stand in front of the fire very long. The fire in the fireplace warms you by a method of transfer of thermal energy called radiation. No substance moves for the transfer of thermal energy by this means to be effective.

Only radiant energy is transferred from the source outward in all directions. When you step out into the sunlight, you feel radiant energy that is radiated from the sun. You have probably noticed how hot objects get if left out in the sun for any length of time. Benjamin Franklin cut small squares of cloth of equal area and placed them in the sun on top of fresh snow. He found that the darker cloth sank deeper into the snow by melting the snow with the thermal energy it had absorbed. It has been found that black bodies are good absorbers of radiant energy. They are also good emitters of radiant energy. You know that heat flows from hot bodies to colder ones. The processes by which thermal energy can flow from hot bodies to colder ones are radiation, convection, and conduction.

You have observed heat appearing when work has been done, such as in the use of simple machines. *Thermodynamics* is the study of quantitative relationships between other forms of energy and thermal energy.

The form of energy involved with simple machines is mechanical energy. Thermodynamics deals with heat and mechanical energy or any other form of energy for that matter. The first law of thermodynamics states that when energy changes form there is no loss of energy. Some of the energy may end up in the form of thermal energy, but the energy does not disappear.

Plain English

Thermodynamics is the study of quantitative relationships between other forms of energy and thermal energy. Literally the word translates as "thermal movement," or "movement of heat."

Heat and Work

I demonstrated the relationship between heat and work to my students using a known source of heat and a Styrofoam cup of water (Styrofoam has about the same specific heat as water). The source of heat was placed in the Styrofoam cup of water for a measured time. The equation that relates the reversible equivalence of work and heat was used to calculate the constant J=4.19 joules/cal. A summary of calculations made is: W=JH, W is the work and H is the amount of heat developed, Pt=J, $\left(m_{cup+water}c_w\right)\left(t_f - t_w\right)$, t_f is the temperature at the end of the time period that the heat source was in contact with the cup of water and t_w is the initial temperature of the cup and water. $J = \dfrac{Pt}{\left(m_{cup+water}\right)c_w\left(t_f - t_w\right)}$, where P is in watts and t in seconds. I demonstrated the trials with the help of students because the source of heat can be very dangerous. I usually prepared three cups of water and after making three trials we found J to be about 4.2 joules/cal. Here the work is done by the source of power for about one minute for each trial.

The quantity of heat resulting from the work is calculated the same way that we have done throughout this chapter. This is just one experiment that shows that when work is done and heat results, no energy is lost. This is consistent with the first law of thermodynamics. You know that in order for this experiment to yield good results the cup and water must have the same temperature as the surrounding atmosphere as their initial temperature. That is why I had the source work for only about a minute so that the heat transferred to the atmosphere would be at a minimum. You already know that the result obtained is good because the goal is scientific truth and not absolute truth. That means that any time an experiment is conducted, error is involved and the careful experimenter attempts to account for as many errors as possible.

The Least You Need to Know

- Explain the difference between temperature and thermal energy.

- Recognize the difference between a change in temperature and a temperature reading.

- Use specific heat to calculate quantity of thermal energy.

- Calculate a quantity of thermal energy required to change the state of a material from a solid to a liquid using the heat of fusion.

- Determine the amount of thermal energy required to change the state of a material from a liquid to a gas using the heat of vaporization.

Chapter 16

Sound Energy

In This Chapter

- ◆ Making sounds
- ◆ A wave model for sound
- ◆ How fast sound travels
- ◆ The speed of sound and temperature
- ◆ Reinforced sound

Hearing is one of the senses that we tend to take for granted. Sound originates in ways other than a singing bird or a barking dog and those who are fortunate to have sensitive ears are able to hear sound.

You have probably heard an echo of your voice when you call a name or greeting loudly across a valley. In this chapter, you will calculate the speed of sound as well as the distance it travels under known conditions. You will consider a model for explaining the behavior of sound and use it to predict when a reinforced sound can be heard under certain conditions. Take the information discussed here and apply it to the practice of using your ability to sense sounds.

Where Does Sound Come From?

Those of us who have healthy ears are grateful to be able to enjoy that special sense of hearing. The energy to which our ears are sensitive is called sound energy or sound. You have probably heard the question about a tree falling in the forest and whether there is sound resulting from the falling tree if no one is there to hear it. The physiologist might tell you that in order to have sound you must have a source, a medium to transmit the sound, and a receiver of sound. He might say since there is no receiver, there is no sound. The physicist might say that sound is a special disturbance of matter to which the ear is sensitive. He might say that sound is there whether it is received by the ear or not. Those same special disturbances may also be beyond the ability of the ear to detect them. The disturbances in matter that we refer to as sound are the topic of this chapter.

As an approach to explaining the disturbances of matter called sound, let's refer back to a topic you considered earlier in this book. Remember the simple pendulum.

You found that once set in motion the pendulum moves with simple harmonic motion. The motion is a vibratory motion that is repeated over and over again. You may want to refer to Chapter 8 and Figure 8.3 to help you to recall information discussed there. One vibration can be traced from equilibrium position to maximum displacement on one side to maximum displacement on the other side and back to the equilibrium position. The maximum displacement from the equilibrium position is the amplitude of simple harmonic motion. One vibration is also called one cycle. The time for the pendulum to complete one cycle is called the period of the motion. The reciprocal of the period is the frequency of the motion. The period is measured in seconds and the frequency in cycles/s or hertz (Hz). The disturbances in matter that are sensitive to our ears closely approximate the to-and-fro motion of the pendulum.

Newton's Figs _____

Any object undergoing simple harmonic motion has associated with it not only a period of motion but also a cycle. One cycle starts at equilibrium, goes to maximum displacement in one direction and back to equilibrium, then to maximum displacement in the other direction and back to equilibrium. Remember that the maximum displacement from equilibrium is called the amplitude.

The strip of metal in Figure 16.1 is secured to a desk so that one end is free to vibrate in a vertical plane.

You have probably placed an object on a table in a similar fashion to that in the diagram and flipped it to set in motion and hear the hum it creates. A plastic ruler works fine if you want to try it. Like the simple pendulum, it vibrates with a motion that closely approximates simple harmonic motion.

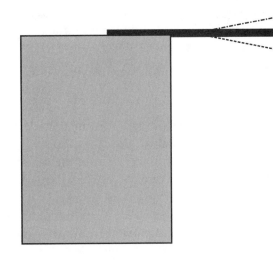

Figure 16.1

The vibrating metal strip compresses air particles on one side and then pulls away from them leaving a void into which the particles move back.

The maximum disturbance is labeled b for bottom and t for top and the equilibrium position is e. When the metal is set in motion, it vibrates from e to t back to e to b and back to e, then repeats the motion over and over again. The path just completed is called one vibration or one cycle. The time to complete one cycle is the period and the reciprocal of the period is the frequency of the motion. This vibrating metal is much like the simple pendulum, but it is different in that you can hear the sound associated with the disturbance of the metal strip.

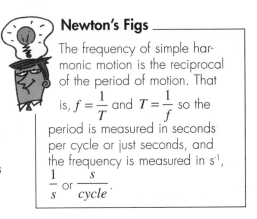

Newton's Figs

The frequency of simple harmonic motion is the reciprocal of the period of motion. That is, $f = \dfrac{1}{T}$ and $T = \dfrac{1}{f}$ so the period is measured in seconds per cycle or just seconds, and the frequency is measured in s⁻¹, $\dfrac{1}{s}$ or $\dfrac{s}{cycle}$.

Sound comes from vibrating matter like the metal strip. The vibrating string of a violin or the booming of a kettledrum are other examples of disturbances of matter that we detect as sound. I enjoy the sound of the plucked strings of a harp. Almost everyone enjoys the vibrating strings of a piano when the tiny hammers set them in motion. The vibrating air columns of the flute and the tuba are pleasing to the ear. The vibrating reed of the saxophone causes air columns with adjustable lengths to vibrate, producing sound that lifts the spirit. You enjoy the sound of the vibrating vocal cords of favorite singers and favorite speakers. The vibrating vocal cords of your cat or dog communicate their needs to you. There are so many vibrating objects of matter that create disturbances that we sense as sound. I have mentioned only those sounds that are pleasing to the ear but you know that we should include things like the buzz of the alarm clock.

The loud sound of a military jet breaking the sound barrier can be frightening. Sound comes from vibrating matter; some vibrations are audible and some are not. You will be able to identify the difference in the two later.

Types of Waves

The vibrating metal strip transfers mechanical energy to the air particles surrounding its motion. As it moves from *e* to *b*, the air particles below the strip are pushed downward, compressing them. That leaves a void in the space above the metal strip. The air particles above the strip are able to move into that space. That increases the volume that they have to occupy and thus decreases their pressure. You know that the strip has its maximum velocity when it passes through *e* so it causes maximum compression of air particles below the strip at that time. While moving from *e* to *b*, the strip continues to compress the particles but the compression becomes less and less as the speed slows to zero at *b*. As the metal moves from *b* to *e* the particles beneath the strip get farther apart.

A Johnnie's Alert

Even though the diagrams of waves are linear, remember that waves can be thought of in two dimensions like water waves or in three dimensions as expanding spherical surfaces.

The particles continue to be separated as the strip moves from *e* to *t*, but they become less and less separated as the velocity slows to zero at *t*. From *t* to *e*, the particles beneath the metal strip are being compressed more and more until maximum compression is reached at *e* when the cycle begins again. A picture can help you to visualize what is going on with a disturbance by the vibrating strip but a review of some notions helps you to interpret the picture.

Transverse Waves

There are many models used in science to explain complex ideas. One model that is of particular interest at this time is the wave. A wave is familiar to you as in the motion of a flag or the motion of the sea. After four years in the Navy, I still get seasick just thinking about the waves on the ocean. There are several types of waves, but we will concentrate on just two of them. The *transverse wave* is probably the most familiar type of wave to you. At some time in your life you have held one end of a rope in your hand and sent a loop down the rope by shaking your hand perpendicular to the rope. The loop traveled down the rope all by itself without having any particles of the rope making the trip with it.

The particles of the rope vibrate up and down, assuming that you created a vertical loop, perpendicular to the rope that determines the line of travel of the wave. A wave is a series of disturbances that travels through a medium because the particles of the medium vibrate. When the particles vibrate perpendicular to the direction the wave is traveling, the wave is called a transverse wave.

Plain English

A **transverse wave** is a series of disturbances traveling through a medium in which particles of the medium vibrate in paths that are perpendicular to the direction of motion of the disturbances of the wave.

Longitudinal Waves

Sound is not thought of as a transverse wave because of the behavior of the particles of the medium. If you think about the vibrating strip of metal, the particles of the medium, the air, are vibrating but vibrating in paths parallel to the direction of the motion of the disturbance. So sound does not fit the transverse wave model. Sound can be thought of as a *longitudinal wave* because of the vibrations of the particles of the medium.

A longitudinal wave is a series of disturbances moving through a medium in which the particles of the medium vibrate in paths that are parallel to the direction of travel of the wave. The loop for the transverse wave and the disturbance created by the part of the vibration of the metal strip moving downward are called pulses. A *pulse* is a single disturbance or a wave of short duration.

The pulse you create in the rope by a flick of the wrist and the pulse the metal strip makes in the downward part of a vibration both travel through the media without having matter move along with it. That is a spectacular thing when you think about it. The pulse in the rope travels through the medium from your hand to the other end without any matter having traveled the space in between. The only thing that travels that distance is the disturbance.

Plain English

A **longitudinal wave** is a series of disturbances traveling through a medium in which the particles vibrate in paths parallel to the direction the disturbances of the wave are traveling. A **pulse** is a wave of short duration.

Properties of Waves

You can model the motion of the metal strip with your hand, and the rope is a record of the motion of your hand. Hold the rope at a position you may call the equilibrium position, then move your hand upward then back downward to equilibrium to create a pulse that has the shape of one of the humps on a camel's back. If you repeat that motion but also continue through the equilibrium position downward and then reverse the motion upward to equilibrium, you create the camel's hump followed by a bowl-shaped pulse. If you continue that motion with a certain frequency for your hand, you create a train of pulses from your hand to the other end of the rope. The train will look like a camel's hump followed by a bowl shape one after the other. Similarly, the vibrating strip will send a train of pulses through the air with regions where the particles are compressed followed by regions where the particles are spread apart.

The picture referred to earlier is in Figure 16.2, where you see a transverse wave and a longitudinal wave diagrammed for comparison.

Figure 16.2

The diagram of the transverse and longitudinal waves displays their distinguishing characteristics.

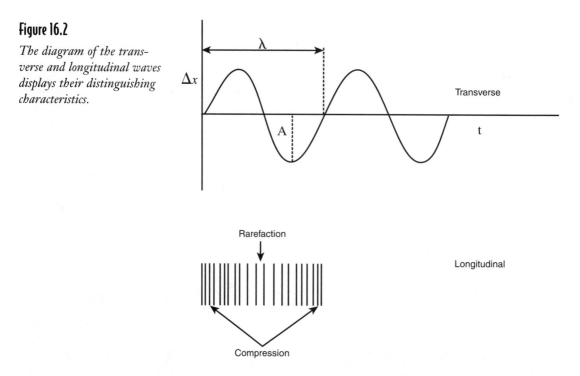

First look at the transverse wave that is a graph of the displacement of the tip of the bottom side of the vibrating metal strip versus the time. The amplitude of the displacement, the maximum displacement of the tip, is labeled A in the diagram. The part that looks like a camel's hump is called a *crest* and the bowl-shaped portion is called a *trough*.

Plain English

The **crest** is that portion of the graph of a transverse wave that lies above the time axis and the **trough** is that portion of the graph of a transverse wave that lies below the horizontal time axis. The **wavelength** of a transverse wave is the distance from the beginning of a crest to the end of an adjacent trough. It can also be thought of as the distance from the point of maximum displacement in one crest to the point of maximum displacement in the next closest crest. One wave is made up of a crest and a trough. The distance from the beginning of a crest to the end of the crest is one half of a wavelength and the trough measures one half of a wavelength. A **compression** is that part of a longitudinal wave where the particles of the medium are pushed closer together. A **rarefaction** is that part of a longitudinal wave where the particles of the medium are spread apart the most.

The distance from the beginning of a crest to the end of an adjacent trough is the *wavelength*, labeled λ in the diagram, of the wave. Did you notice that the horizontal axis is time and not distance? You are very observant. The time to complete one full wave is what I have labeled λ on the graph. The relationship that is implied here becomes clear as we develop this idea.

The longitudinal wave is created by the motion of the air particles near the bottom of the metal strip. When the metal strip is passing through the equilibrium position in Figure 16.1 moving downward, a *compression* is being created.

As the strip continues down to b in that diagram, the particles are getting farther apart because the bottom of the metal is moving away from them. When the strip continues from b and passes through e, the particles are at a maximum distance apart in the part of the longitudinal wave called a *rarefaction*.

Continuing the motion of the strip toward t, the strip is slowing down so the particles near the bottom of the strip are not as far apart as they were. Traveling from t downward the strip is speeding up, causing the particles beneath the metal to compress and reaching maximum compression at e as the strip continues its downward motion. You have just traced the creation of one complete longitudinal wave that starts with a compression and ends with the next compression in a train of waves. That complete wave-pulse has a length that is called the wavelength of a longitudinal wave. It would be the distance that the beginning of the pulse moves out from the source during the time that it takes for the whole cycle to occur.

Stating that sound is a longitudinal wave means that the longitudinal wave is the best model we presently have for explaining the behavior of sound. Scientists often understand a model and borrow it to see how well observed phenomena fit the model.

Newton's Figs _____

The compressions and rarefactions of a longitudinal wave are regions where particles are closer together and farther apart, respectively. The action of pushing and spreading apart throughout the room from the source of sound is the wave nature of sound. It spreads like ripples on the surface of water (but in three dimensions) and remains circular because the speed of the wave's disturbances are constant.

There are several sites on the web where you can find dynamic models that illustrate this motion. You might enjoy finding some to get a moving picture of what is described in this chapter.

If there is a good fit, then the model is used to explain the observations of a material or event. If the model not only explains most observations about a phenomenon but also predicts other things that an experimenter might look for relating to his other observations, it is a model used widely. If the predictions lead to understanding the phenomenon more, the model becomes the skeleton of a healthy body of knowledge. That means that viewing sound as a wave, and not only that but a longitudinal wave, lends itself to a better understanding of sound.

Speed of Sound

The wave model provides for several properties of sound. Sound requires a medium for transmission. We hear each other talking by the transmission of sound through air. If you dive to the bottom of a swimming pool during summer fun, you can hear an object scrape the side of the pool. Water is actually a better medium for the transmission of sound than air.

You have probably played with a tin can that has a string through a hole in the bottom. A sibling or friend talks to you by speaking into another tin can at the other end of the taut string. The voice is transmitted very well over the string, but the arrangement works even better if the string is replaced by wire. Solids are also a good medium for sound transmission.

Newton's Figs _____

Outer space like the region above the surface of the moon is about as close to a vacuum as you can get. Remember that astronauts dropped a feather and a hammer and both hit the surface of the moon at the same time. The hammer and feather experienced no opposition to their motion from air as they fell to the moon's surface.

Is there anything that does not transmit sound? Well, the closer you can get to nothing, a vacuum, the closer you get to a medium that does not conduct sound. Experiments have been done in which air is removed from a space containing a bell or some other noisemaker. It is found that the closer the space is to a vacuum, the more difficult it is to detect any sound. Sound requires a medium for transmission.

You are well aware that sound does not travel as fast as light. Sound does have a finite speed as does light, but over short distances the speed of light is so fast that it seems practically instantaneous.

Motion and Sound

You have probably observed the delayed sound of a distant shovel on a cold winter morning? You see the action and later hear the result of the action. You see the flash of lightning in a thunderstorm and hear the thunder a short time later. Sound does travel with a definite speed that is consistent with the wave model. The speed of sound is about 331.5 m/s at 0°C; that is about 1087 ft/s at 32°F or about 740 mi/hr. I know that you can convert these quantities from one system to the other, but this may save you a little time for the next idea.

What type of motion is involved in the description of sound or sound waves? Uniform motion, of course. The speed is constant, and the uniform motion is described simply using the equation distance is equal to rate multiplied by time.

If you are caught out in a thunderstorm and see a flash of lightning, start counting one-thousand-one, and so on, and when you hear the thunder, mentally multiply the number you counted by 1100 ft/s to get a good idea of how far away the storm is from you! You may have this figured out already in a form like this: five counts about a mile, three counts about half a mile, one count right on top of you! If it is one or two counts, take cover quickly—but not under a tree, please! Later you will find that a tree is a prime target for lightning at any time. Did you notice that we are measuring distance by using time again? Good for you! The implication is that the object of the discussion is traveling at a constant speed. Sure enough, the sound wave is traveling at a constant speed.

Wavelength and Frequency

Suppose we pursue the notion of a sound wave just a bit further. Remember that I suggested that your hand can model the vibrating metal strip if you do several things, one of which was to set your hand in motion at a certain frequency. What do you suppose would happen if you doubled the frequency of your hand on the rope while keeping all other conditions the same? This means you have to maintain the same amplitude and keep the rope at the same tension. The wavelength in the rope will automatically get shorter? You bet! If you continue to increase the frequency of your hand, you are apt to faint if you are not careful because that is work. The greater the amplitude, the greater the work you have to do.

I used a small-diameter spring to demonstrate this situation for my classes, and at high frequencies my face was red, much to the delight of my students. The strength of my arm was gone by the end of the trial. What would you expect of your graph if you were to plot the wavelength versus the frequency? Applying graphical analysis techniques used in Chapter 5, I plotted this for you in Figure 16.3. Follow the line of reasoning using graphical analysis carefully as you may use the results in discussions involving the behavior of waves.

Figure 16.3

The graph of wavelength versus frequency suggests an inverse relationship.

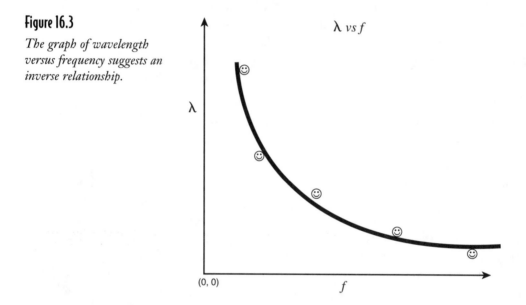

Notice that graph suggests an inverse variation, that is, $\lambda \propto \dfrac{1}{f}$. I graphed that for you in Figure 16.4, where you can see that the graph confirms the notion because the straight line means that $\lambda = \dfrac{k}{f}$ or $f\lambda = k$ where k is the constant of proportionality. Can you guess what that constant might represent? What was it that was staying constant?

Remember that k is the slope of the straight line for that graph. What do you suppose that constant k is? You can get a hint by checking the units of the slope if you have not already guessed what it is. The units of k turn out to be $\dfrac{meters}{\frac{1}{\frac{1}{s}}}$ and that turns out to be $\dfrac{m}{s}$, the units of speed! The graphical analysis of the hypothetical data suggests that the speed of the wave in the medium is equal to the product of the frequency of the wave and the wavelength of the wave, $v = f\lambda$. The relationship is one that you can expect any time you are discussing waves or the wave nature of physical phenomena. Try a few problems to make these ideas a part of your experience.

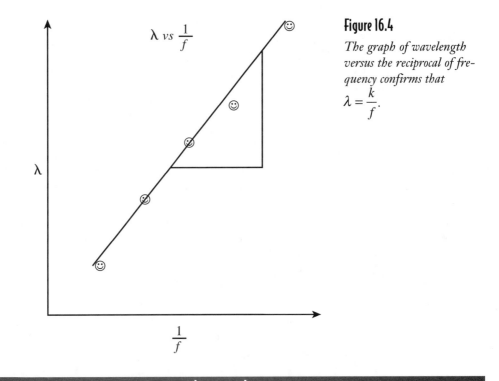

Figure 16.4

The graph of wavelength versus the reciprocal of frequency confirms that

$$\lambda = \frac{k}{f}.$$

Physics Phun

1. A source attached to a rope vibrates at 45 cycles/s, 45 Hz, and creates waves in the rope traveling at 15 m/s. What is the wavelength and period of the waves in the rope?

2. A train of waves moves along a wire with a speed of 25 ft/s. What is the frequency and period of the waves if they are 5.0 in length?

3. Audible sound ranges from about 20 cycles/s, or 20 Hz, to 20,000 Hz. What are the wavelengths for these frequencies? Note: Vibrations below 20 Hz and above 20,000 Hz are not detectable by the human ear.

Effects of Temperature on Sound

You anticipated that the temperature plays a role in the behavior of sound since the speed of sound was given at $0°C$. At a given temperature the speed of sound is constant in a given medium. Even though you may have available a solid, a liquid, and air at the same temperature, the speed of sound is constant in each medium but the speed has a different value in each medium.

A Johnnie's Alert

Like other phenomena that can be described as waves, sound waves diffract (bend around corners), refract (change direction when traveling from one medium to another), and reflect (bounce off surfaces).

Plain English

The **Doppler effect** is the apparent shift in the pitch of a source of sound because of the relative motion between the source and the observer.

Consider some happenings involving sound that you have observed at a given temperature. Listening to a band play on the football field in the evening you notice that you can hear all of the different instruments at the same time. You hear high notes and low ones. The pitch you hear depends on the frequency of the vibrating reed or air column; the higher the frequency, the higher the pitch. Since all of the notes from the various instruments arrive at your ear at the same time, all frequencies must travel at the same speed. Recall that since the speed is constant, each frequency has associated with it a definite wavelength as we found in the last section.

Remember standing at a train station when a speeding train passes by with horn blaring. You probably noticed the warning horn at some distance from the station. When the train passed the station, the pitch of the horn seemed to suddenly change to a lower tone. That phenomenon is called the *Doppler effect*.

Even though the train is sounding the same tone at all times, when it is coming toward you the frequency appears higher than it does when the train is standing still at the station. Recall that the source of sound sends out a series of waves, one after the other. When the locomotive is coming toward you, you are receiving more waves each second than you do when the train is standing still. Even though the speed of sound is constant, you are observing the speed of sound plus the speed of the train for the waves you observe with your ears.

As the train passes you, the observation you make is the difference in the speed of sound and the speed of the train. Therefore your ears perceive fewer waves each second, resulting in a lower frequency and lower pitch. The apparent shift in pitch as a source of sound passes you is the Doppler effect.

Reflection of Sound

An echo is another event that you have observed at some time. The echo is a reflection of sound from a surface. Sound reflects from a surface very much like a ball bounces off a surface. Your observation of the reflection of sound in most cases has probably been a novelty like hearing your voice calling back from the bank of a distant canyon. Reflections of sound can be troublesome if you are sitting in the wrong place in an auditorium that does not have good *acoustics*.

If the acoustics of a room are bad, reflections of sound from walls or other surfaces can meet in such a way as to cancel sound altogether. You tend to think of a sound wave as linear but it can also be compared to a ripple of a water wave. Actually a sound wave is three-dimensional, spreading away from a source of sound in the shape of a sphere. If two sound waves meet at a point so that a rarefaction from one source meets a compression from another source of the same frequency, they will completely cancel each other. If your ear is at that point it detects no sound!

Plain English

Acoustics are the sound-producing qualities of a room or auditorium.

Two sources of sound of the same frequency generating sound waves at exactly the same time can create some fantastic observable patterns. Some young physicists set up two sources of sound in an arena and invited my class out to a local college to observe what happens under these conditions. Since sound waves are three-dimensional, instead of canceling at a point, they actually cancel sound in regions of space.

My students walked around the arena, climbed to different levels, and observed absence of sound separated by regions of reinforced sound both vertically and horizontally. You can observe the same thing on a limited scale if you buy or borrow a tuning fork with a frequency of one thousand cycles/s or more and do the following.

A Johnnie's Alert

Vibrating sources of sound have images in reflecting surfaces just like you have an image in a mirror. Points where waves cancel are called nodal points. Two-dimensional waves like water waves cancel in lines called nodal lines. Waves in three dimensions cancel in planes called nodal planes. Nodal lines are hyperbolas and nodal planes are hyperboloids. Special conditions must exist for cancellation to take place.

Have a friend stand by a flat, smooth wall or surface, start the tuning fork, and hold it 30 to 50 centimeters from the flat surface. You should stand by the same surface about 10 meters away from the vibrating tuning fork. With your back against the wall, walk slowly away from the wall and you will observe regions of silence and regions of reinforced sound. After you are convinced of your observations, you can observe the same thing at different levels above the floor.

Interference of Sound

You thought we needed two sources of the same frequency generating sound at exactly the same time? That is a requirement, and you have a fantastic sense of details. The tuning fork has an image exactly like itself in the wall or surface. Just like you have an image of yourself in the mirror, but that is a topic to spend time on in Chapter 20. The image of the tuning fork is the same distance in back of the wall as the tuning fork is in front. That means we have the condition for this phenomenon to work. The thing you observe is called an interference pattern and is characteristic of all waves under the same conditions. The interference pattern is made of areas of constructive interference where the sound is louder and destructive interference where the sound is softer.

All of these things occur because sound is a wave and the speed of sound is constant for a given temperature. The speed of sound changes with temperature. The speed of sound in air is about 331.5 m/s at 0°C and increases about 0.60 m/s for every degree Celsius increase in temperature. In the FPS system, the speed is about 1087 ft/s at 32°F and increases 1.1 ft/s for every degree Fahrenheit increase in temperature. Since sound travels faster in air at warmer temperatures, for a given frequency the wavelength is slightly longer than at freezing temperatures.

Suppose the temperature is 70°F; then the speed of sound is 1087 ft/s plus (70-32)F° (1.1 ft/s), which turns out to be about 1129 ft/s. For a frequency of 256 cycles/s, the wavelength is about 4.2 ft at 32°F and about 4.4 ft at 70°F. Remember that even though the speed of sound increases with temperature, all of the phenomena discussed are still observed at a given temperature. Notice that the frequency of the sound did not change due to the change in temperature. It is determined by the frequency of the source. In this case that would be the tuning fork that was used.

Newton's Figs

The speed of sound is less at temperatures below 0°C or 32°F than it is at those temperatures. It is greater at temperatures above 0°C or 32°F than it is at those temperatures.

Physics Phun

1. How much time is required for sound to travel 6.78 km when the temperature is 5°C?

2. How many feet long is the wavelength of a wave generated by a tuning fork with a frequency of 256 cycles/s if the temperature is 86°F?

3. A man drops a stone into a well 155 m deep. The temperature is 8°C. How long does it take for him to hear the stone splash in the water below?

Resonance

You have helped a small child or a friend enjoy the swings in the park. Standing in back of the person in the swing and giving them a push in the swing at just the right time causes them to go higher. The squeals of delight from a small child encourage you to continue giving just the right push at just the right time. The push at just the right time to get the swing and happy cargo vibrating at the same frequency as the push is called *resonance*. Energy is absorbed by a system in harmonic motion if the energy is input at the same frequency.

That is why you can't just sit in a swing and fling your legs around in any old rhythm. You time your push so that the swing is at the same place and traveling in the same direction at the time of the push. The condition of resonance can be quite dramatic at times. The Tacoma Narrows Bridge received pushes from the wind at just the right frequency for it to swing so wildly that it snapped and fell into the water. Some say that there are singers with voices that can break a glass when they hold a sustained note that has the same natural frequency as the glass.

Plain English

Resonance in sound is the increased amplitude of vibration of an object caused by a source of sound that has the same natural frequency as the object's frequency. If two vibrating objects have the same natural frequency, one of the objects can cause the other to vibrate at its natural frequency. A person singing near a piano can cause strings in the piano to vibrate sympathetically or in resonance with their voice.

Observing Resonance of Sound

The speed of sound can be measured by establishing resonance in closed and open tubes. Have you seen the big pipes behind the organ in a church? Why are they all different lengths? Do you think some of them might have a natural frequency associated with them? Read on .... The resonance of the air column in the tube, with a vibrating source of sound, is observed as a louder sound created by the correct length of vibrating column of air in a tube.

It is found that sound (an example of a longitudinal wave) incident on a surface of a new medium can reflect from the new medium in one of two ways. If the new medium is more dense than the original medium that the wave is located in, a compression is reflected at the surface as a compression and a rarefaction is reflected as a rarefaction. If the surface is less dense, such as leaving cooler air and going into warmer air, a compression is reflected as a rarefaction and a rarefaction is reflected as a compression. Refer to Figure 16.5 as the reflections of a compression from each end of the tube are traced and the distance traveled recorded. This idea of reflection is common to all waves, but reversal ideas upon reflections will be different for transverse waves. Watch for that.

Figure 16.5

The conditions for resonance in a closed tube are shown by following the reflections of a compression.

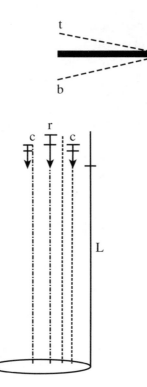

Resonance of a Closed Tube

The vibrating source of sound is a tuning fork held above a tube. The correct length of an air column can be established experimentally by holding an open tube with one end in water. Hold the tuning fork just above the mouth of the open end and then raise and lower the other end of the tube in the water until you hear an obvious increase in volume of sound. The increased volume of sound signals resonance; the column of air is vibrating at the fundamental frequency of the source or it is vibrating at some harmonic frequency of the source. Follow these steps to see how resonance in the closed tube is explained.

Step 1: Assume that the vibrating source has completed a compression when the tip of the lower side of the source is at position *b* in Figure 16.5. A compression is sent down to the bottom at that time. In order to have resonance for the shortest length of the closed tube, the source must be creating a compression at the same time a reflection from the air at the top of the tube is a compression.

Step 2: The compression travels to the bottom of the tube and is reflected as a compression.

Step 3: The reflected compression travels to the top of the tube and is reflected as a rarefaction as shown. At this time, half a vibration of the source of vibration, the sound has traveled the length of the closed tube twice.

Step 4: The rarefaction travels to the bottom of the closed tube where it is reflected as a rarefaction.

A Johnnie's Alert

Reflection of sound can be helpful as in the vibration of a resonant air column or harmful as in the case in which sound waves cancel causing you to miss some important lines in a play.

Step 5: The reflected rarefaction travels back up the tube and is reflected at the top as a compression. At this time, second half of a vibration of the vibrating source, the sound has traveled the length of the closed tube twice.

Step 6: At the instant the reflected compression is created and just starts its trip downward, the source must have created another compression that is traveling in the same direction at the same time in order to have resonance. That is, in order to receive just the right push at just the right time in just the right direction, the two compressions must satisfy all of these conditions.

Step 7: The source will have completed one cycle to meet these conditions. That means the distance the sound has traveled in one cycle of the vibrating source is one wavelength, four times the length of the closed tube. The length of the closed tube is $\frac{\lambda}{4}$, which is one fourth of the wavelength of the vibrating source.

Notice that the conditions for resonance as outlined here are for the shortest length of the closed tube.

By varying the length of the closed tube, conditions for resonance can be established for the harmonics of the vibrating source. The lengths of a closed tube for the fundamental frequency and its harmonics are $\frac{\lambda}{4}, \frac{3\lambda}{4}, \frac{5\lambda}{4}$ ....

Newton's Figs

The shortest length of a closed tube that resonates with the source of sound is one quarter the wavelength associated with the fundamental frequency of the source.

The wavelength for the fundamental frequency is about four times the length of the tube or $\lambda = 4L$. The diameter of the tube does make a difference for the calculation of λ, so to make the calculation more accurate use $\lambda = 4(L + 0.4d)$ where d is the diameter of the tube. The tube can be glass, metal, or even a cardboard mailing tube. You must try this experiment and compare your calculation of the speed of sound with the value you calculate for the temperature that day. That is, since $V = f\lambda$ as determined in an earlier experiment, using the results of the resonance experiment $V = 4f(L + 0.4d)$. Compare that value with $V = (331.5 + 0.6\Delta t) \, \text{m/s}$ or $V = (1087 + 1.1\Delta t) \, \text{ft/s}$.

The Least You Need to Know

◆ Identify sources of sound.

◆ Describe the features of transverse waves and longitudinal waves.

◆ Explain the relationship of frequency and wavelength with the velocity of sound.

◆ Calculate the velocity of sound in air at many Celsius or Fahrenheit temperatures.

◆ Use the reflection of sound waves to explain resonance in sound.

Part 5

Tiny Things Like Electrons in Motion

The last leg of the journey leads into the world of tiny particles of matter. Tiny particles that have such provocative names as atoms, molecules, alpha particles, and protons are the characters crowding the streets of the area. You will also get a glimpse of the results of even smaller particles bouncing off mirrors or forming images by passing through a double convex lens. Other tiny particles work to light up reading lamps and help to cook your morning toast.

Chapter 17

Atoms, Ions, and Isotopes

In This Chapter

- ◆ An idea of what the atom looks like
- ◆ Ions from atoms
- ◆ Same family, different name
- ◆ Atoms and molecules
- ◆ Atoms and their parts in motion

The atom and molecule are discussed as the building blocks of matter. You will find that there is much more to the story of what matter is made of as you explore particles. There are several places you may want to visit on the Internet to find out more about the many particles with strange-sounding names. Here are some sites (there are many more): http://particleadventure .org, http://whatisit.techtarget.com, and www.chemtutor.com.

A brief visit to any one of these sites and you will see that beginning our discussion with molecules, atoms, neutrons, protons, and electrons is about all that we can handle in this chapter. You may choose to do a lot more exploring but we will begin with some of the simple things.

A Model of the Structure of the Atom

Imagine that you travel from your frame of reference into the insides of matter. The first particle you see is fairly large as these tiny particles go. It is called a *molecule*.

You may see molecules like hydrogen, water, or sugar. The hydrogen molecule is fairly small, a water molecule a little larger, and as molecules go, a sugar molecule is huge. These are the types of particles we were referring to earlier when we talked about the behavior of solids, liquids, and gases. Prying a little deeper into matter, you find tiny particles called *atoms*.

Atoms Next

That is a good place to stop for the moment so that we can take a good look at the atom. It is an interesting world at the level of the atom. How about recording a couple of measurements? Length for instance: The hydrogen atom is about $0.6 \overset{\circ}{A}$ in diameter. The largest atoms are around $5 \overset{\circ}{A}$ in diameter. $1 \overset{\circ}{A}$ or $1 \overset{\circ}{A}ngström = 1 \times 10^{-10} m = 1 \times 10^{-1} nanometers$. Obviously, you would not use a meter stick to make those measurements. The mass of the hydrogen atom is about $1.67353 \times 10^{-27} kg$, so a triple beam balance would not be of much use on this level either. These measurements require much more physics than we cover here but you might be interested in researching procedures used to determine these values. For our purposes we are satisfied to use the values given.

Plain English

A **molecule** is the smallest particle of a compound. An **atom** is the smallest particle of an element that exists alone or in combinations. An **element** is a substance like copper, hydrogen, or neon that cannot, by chemical means, be broken down into other substances.

A Johnnie's Alert

It is important to be familiar with both the nanometer and the angstrom because both are used in reference books that you might read. 1 angstrom = $10^{-10} m = 10^{-8} cm = 0.1 nanometers.$

We talk about these particles as if they have always been part of the daily experience. Scientists have developed most of the information known today about atoms since the latter part of the nineteenth century. Three models of the atom will be discussed briefly to provide a feeling about the evolving nature of how we think about atoms. Thomson associated a negative charge, an idea developed in a later chapter, with the electron, a constituent of the atom. He proposed a model that assumed there was an equal amount of positive charge associated with the atom, now known as protons. Bohr's idea about shells for the location of electrons was helpful, but his ideas came after Rutherford's discovery of the positively charged nucleus.

These two ideas improved Thomson's original model. Since electrons are attracted to the positive nucleus, there is a problem with the fact that electrons were not spiraling into the nucleus of the Rutherford model. Later theories were developed to try to explain why accelerating electrons were not continually giving off radiant energy. Rutherford's contribution of the nucleus containing positive charge and most of the mass of the atom did, however, present an improved model over earlier ones.

The Bohr model used the nucleus of the Rutherford atom and proposed discrete orbits for the electrons to remedy the problem of having them spiral into the nucleus as they lost energy. His model provided for the electrons to jump into an orbit with higher energy by absorbing a quantum of energy. The electron gives up or emits a quantum of energy if it falls back into a more normal orbit. You can see that man has continued to work on his interpretation of the particles of matter by improving the model of the atom. We consider the atom as a combination of these models along with some ideas of a current model. You might think of the atom as having a definite nucleus, and the electrons orbiting the nucleus, not in discrete orbits, but rather in nebulous areas more like a cloud.

Parts of the Atom

The models of atoms are discussed as if you know a lot about the atom already, and that is no doubt true. We need to begin with a model and define the parts as we go. Start with the *nucleus*. Since Rutherford's research proposed the nucleus, it has been a part of the model of an atom. In his experiment with *alpha particles*, which will be discussed later, he found that most of the atom is empty space except for the nucleus.

Plain English

The **nucleus** of an atom is considered the core of the atom and is made up of protons and neutrons. The **alpha particle** is the special name given to the nucleus of the helium atom. In other words, a helium atom minus its electrons. The **coulomb** is the unit of electrical charge. The **proton** is a positively charged particle with a mass of about $1.7 \times 10^{-27} kg$. The **neutron** is a particle with no charge and about the same mass as the proton. The **electron** is a particle that has a charge of $1.6 \times 10^{19} coulomb$ and a mass of $9.1 \times 10^{-31} kg$.

The nucleus of the atom is made up of *protons* and *neutrons*. The proton has a positive charge. In fact, the value of its charge has the same magnitude as the charge on the electron. The charge on the proton is 1.6×10^{19} *coulomb*, and its mass is about 1.7×10^{-27} *kg*. The neutron is neutral; that is, it has no charge and a mass that is close to the value of the proton's mass.

The protons and neutrons in the nucleus are referred to collectively as *nucleons*. The number of neutrons and protons in the nucleus enable us to identify different kinds of atoms as well as to specify characteristics of a given atom.

As was mentioned earlier, most of the atom is empty space. The atom does have the *electrons* outside the nucleus. Different models suggest that they are in orbits, shells, or clouds. In fact, there are just as many electrons outside the nucleus as there are protons inside the nucleus. The electron is a particle with a mass of 9.1×10^{-31} *kg* and a charge of 1.6×10^{19} *coulomb*. Note that the value of the charge is the same as the proton; however, it is opposite in nature.

The mass of the electron is less than 0.001 of the mass of the smallest atom. That fact leads to the conclusion that the electron is part of an atom and not an atom itself. The charge on the electron was established by Millikan's oil drop experiment, in which he also established the fact that the charge on any charged object was some integral multiple of the charge on the electron. That means that charge is grainy, with grains of definite size. The model of the atom is characterized by a structure in which there is a central region called the nucleus. It is made up of protons and neutrons. There are the same number of electrons outside the nucleus as there are protons that are in the nucleus.

A Johnnie's Alert

Stating that the atom is mostly empty space emphasizes that the nucleus is only about 10^{-4} of the diameter of the atom. Think of it this way. Perhaps you have or had a dorm room. When you were in the room, you occupied a great deal more space than your body actually measures. In the same way, an atom takes up lots of space because of the electrons that are outside of the nucleus. The volume of the nucleus is only a part of the whole volume of the atom.

Isotopes and the Same Element

The model of the atom used here has two important numbers associated with it: the *atomic mass number* designated by the symbol A and the *atomic number* that identifies the number of protons in the nucleus, designated by the letter Z. These numbers are used to identify the element and to indicate the number of nucleons.

Suppose that you are given an atom with 4 protons and 12 nucleons. That means Z = 4 and A = 12, and since A is the number of nucleons then A–Z= the number of neutrons in the atom. The atom has 12 nucleons made up of 4 protons and 8 neutrons. The symbol for an atom is written $^{12}_{4}Y$ where Y is the symbol of the element and can be found on a periodic table of elements. In general, the symbol for an atom is written $^{A}_{Z}Y$. As you can see, Z identifies the element even though the symbol for the element is included in the notation.

Plain English

The **atomic mass number** of an atom is the number of nucleons the atom contains. The **atomic number** is the number of protons in the nucleus of an atom. **Isotopes** are atoms of the same element that have different numbers of neutrons. The **unified atomic mass unit**, u, is the unit of mass used to stipulate nuclear masses.

There are atoms of the same element that have a different number of neutrons. For example, all oxygen atoms have 8 protons, but they can have a different number of neutrons. $^{16}_{8}O$, $^{17}_{8}O$, and $^{18}_{8}O$ are all examples of atoms of oxygen. Nuclei having the same number of protons but a different number of neutrons are called *isotopes*. The oxygen isotopes listed are stable isotopes of oxygen.

All isotopes of an element are not equally abundant. Some isotopes can only be produced in the laboratory. Many isotopes occur naturally, such as the abundant isotope of carbon, carbon twelve or $^{12}_{6}C$. You probably know that the isotope $^{14}_{6}C$ is used for carbon dating. No, not that kind of dating; it is the process of checking the age of ancient objects that were alive at one time.

The unit of measurement for nuclear mass is the *unified atomic mass* unit or u. Carbon twelve, $^{12}_{6}C$, was given the exact value of 12.000000 u on this scale. On this scale, a proton has an atomic mass of 1.007276 u, a neutron has a mass of 1.008665 u, and a hydrogen atom has an atomic mass of 1.007825 u.

You may use your knowledge of unit analysis to define the mass in grams of 1 u. Earlier we found that a hydrogen atom has a mass of $1.67353 \times 10^{-27} kg$ and a mass if 1.00725 u. Write that as a defining equation, then divide both members by 1.00725 to find 1 u.

$$1.67353 \times 10^{-27} kg = 1.00725u$$

$$1.66054 \times 10^{-27} kg = 1u$$

You may use this last result to express nuclear mass in kg given the nuclear mass in u. The atomic mass number is the whole number closest to the atomic mass of an atom. That is, given that the atomic mass of oxygen is 15.9994 u then the atomic mass number is 16. That means that this isotope of oxygen has 16 nucleons, and since the atomic number of oxygen is 8, there are 8 protons and 8 neutrons in the nucleus. Work the example yourself before you look at my solution.

Problem:

The periodic table gives the following information about an element: The symbol is Fe, the atomic number is 26, and the atomic mass is 55.847. What is the mass number of iron symbol Fe? Find the number of protons and neutrons in the nucleus.

Solution:

The atomic number, 26, is given, and that is the number of protons in the nucleus. The mass number is 56 because that is the number nearest the atomic mass of the element. That means there are 30 neutrons in the nucleus because the sum of the neutrons and protons must be 56.

Plain English

A **mole** is Avogadro's number of items. Another way of defining it is a mole is a basic unit of quantity.

You can include more information about the iron atom because you know the mass of u in kg. The mass of an iron atom in kg is given by:
$(55.847)(1.66054 \times 10^{-27} kg) = 9.2736 \times 10^{-26} kg$.

Instead of working with the mass of one atom, many times the scientist deals with the mass of a *mole* of atoms. A mole of anything is Avogadro's number of the items.

Avogadro's number is 6.02×10^{23}. The mass of a mole of a substance turns out to be numerically equivalent to the atomic mass of the smallest particle of the substance. The mass of a mole of iron in grams, assuming that the smallest particle of iron is one atom, is 56g. You may check our previous answer by dividing the mass of a mole by Avogadro's number: $\dfrac{.056kg}{6.02 \times 10^{23}} = 9.3023 \times 10^{-26} kg$, which is essentially the same as before when we used not the atomic mass number but the atomic mass of iron. It is instructive for you to try some of these problems on your own.

Note: Tables usually express the atomic mass of an atom to six significant figures. You are not to be confused by the precision of that information. Use a calculator and express your answers to reflect the same precision.

Physics Phun

1. The atomic number of an element is 47 and the symbol is Ag (silver). What is the mass number if the atomic mass is 107.868 u? How many neutrons and protons are in the nucleus? What is the mass in grams of a mole of silver given that the atom is the smallest particle of silver? What is the mass in grams of an atom of silver?

2. What is the atomic number of a lead (symbol Pb) atom if it is made up of 82 protons, 125 neutrons, and 82 electrons? What is the mass number of the atom? What is the mass in grams of a mole of these atoms? What is the mass in grams of one of these atoms?

3. What particles, and how many of each, make up an atom of gold (symbol Au), atomic number 79, mass number 197?

The Formation of Ions

The atomic model discussed here features a nucleus in the center where most of the mass is found surrounded by mostly empty space except for a cloud of electrons. The electrons occupy regions of space called energy levels. Only a certain number of electrons can be in a particular energy level. All of the energy levels that have any electrons in them are filled for some atoms like argon, helium, krypton, and neon. These elements are stable; that means they do not combine with other elements. They are found on the far right side of the periodic table.

Some atoms have outer energy levels that are incomplete. That may mean that the outer energy level of that type of atom may contain eight electrons, but that particular atom only has six in its outer energy level. It may mean that that type of atom has all its inner energy levels filled, but the outer energy level has only one electron in it even though it needs eight to complete its complement of electrons to become stable. While eight is not a magic number, it is the number of electrons many atoms require to fill their outer energy level for stability. Not atoms like hydrogen or helium, of course, since they only have one or two electrons. Electrons occupying an incomplete outer energy level are called *valence electrons*. The valence electrons are the outer electrons that largely determine the chemical nature of an element.

 Plain English _____

A **valence electron** is an electron in an incomplete energy level. Some atoms can lose their valence electrons and become stable. However, for different kinds of atoms stability is acquired by gaining electrons until the total number of outer electrons is eight. These loosely bound electrons are largely responsible for the chemical behavior of an element. An **ion** is a charged particle that is the result of an atom gaining or losing electrons to exhibit an excess or a deficiency of electrons.

If an atom gives up a valence electron for some reason, since it has complete energy levels in its core, it may become a stable ion. It is no longer a neutral atom, but has become a positively charged particle called an *ion* because it now has one more proton than it has electrons. If an atom gains an electron for some reason, it becomes a stable negatively charged particle because it has one more electron than protons. Neutral atoms become negative ions if they gain electrons. Note: Even though an alpha particle is a positive ion, it quickly acquires two electrons from surrounding matter and becomes a neutral helium atom.

Another Source of Ions

An ion that might be familiar to you is an alpha particle. An alpha particle is the nucleus of the helium atom. Helium has atomic number 2 and mass number 4 so you know quite a bit about this guy given that information. You know that the nucleus has 2 protons and 2 neutrons. Since an alpha particle $_2^4He$ is just the nucleus of the helium atom, it must have a positive charge of plus 2. The alpha particle was mentioned earlier and alluded to this discussion.

Some types of atoms are radioactive. That means that the nucleus spontaneously disintegrates and emits *alpha particles* $_2^4He$, *beta particles* $_{-1}^0e$, or *gamma rays*. Radium $_{88}^{226}Ra$ is a radioactive element that can cause cancer if the human body is exposed to it for long.

Plain English _____

Alpha particles $_2^4He$ are helium nuclei that have a positive charge and are emitted from the nucleus of some radioactive elements like radium. **Beta particles** are high-speed, high-energy electrons emitted from the nucleus of some radioactive elements, and they have a negative charge just like the electrons found in the energy levels around the nucleus. **Gamma rays** are very penetrating, short-wavelength electromagnetic radiation emitted from the nucleus of some radioactive elements. They have no charge, but do carry away energy from the nucleus.

Uses of Radioactive Particles

Radium has been very useful in the treatment of cancer as well as a research tool. It is used in the laboratory as a source of alpha particles, among other things. Rutherford used radium in his experiment in 1911 when he discovered the nucleus of the atom. He used radium embedded in lead $_{82}^{207}Pb$. Since lead absorbs radioactive particles, he left a small opening through which alpha particles traveled to bombard very thin gold $_{79}^{197}Au$ foil. Rutherford and his assistants found that most of the alpha particles passed right through the gold foil, indicating that the foil is very porous to alpha particles. This meant that many of the atoms of gold must have been empty space. Otherwise, the alpha particles would have been pushed from their paths by the charged particles within the gold atoms.

Much to his surprise, a significant number of the alpha particles were deflected at large angles to their paths. Since the alpha particles travel at about 20,000 mi/s, that result suggested that the alpha particles were interacting with something charged positively, the same as his alpha particles. It also suggested that the positive charge of the gold atoms had to be in a relatively small volume. This led to the idea of a positive nucleus for the atom.

CAUTION

A Johnnie's Alert

Particles emitted from radioactive atoms travel at very high speeds but do not travel very far. A sheet or two of printer paper stops alpha particles. A beta particle can travel for several meters in the atmosphere before being stopped but is absorbed by about a centimeter of aluminum. Gamma rays are not charged and travel very fast. They are much more difficult to stop.

Almost unbelievable to Rutherford was the result that some of the particles bounced straight back, opposite the direction they were traveling initially. You remember the interaction of the billiard balls earlier in this book, and you know it would be surprising to see a billiard ball bouncing straight back from a collision. Unlike the billiard balls that touched when they collided, the alpha particles must have collided with something very massive and charged positively. This confirmed the ideas suggested about the nucleus.

Rutherford called the thing that the alpha particles interacted with the *atomic nucleus*. Because the interactions were infrequent, he concluded that the nucleus was very small. The diameter of an atomic nucleus is about $10^{-14}m$, which is the order of magnitude of the diameter of an electron.

Just think about it for a moment: an alpha particle $_2^4He$, traveling at 20,000 mi/s, colliding with the nucleus of a gold $_{79}^{197}Au$ atom!

No wonder there was such a change in direction for the alpha particles. Of course Rutherford did not know their relative sizes then, but his experiment laid the groundwork for our level of understanding of the nucleus of the atom.

The Formation of Molecules

Alpha particles are not the only ions you know about. Some ions are very familiar to you, especially if you enjoy good food. Sodium $_{11}^{23}Na$ has a single valence electron in its outer energy level that the sodium atom is just itching to give up. If the outer electron left, the sodium atom would then have eight electrons just like the noble gases found at the far right of the periodic table. Remember how stable they are. Chlorine is lacking one electron in its outer energy level and would be delighted to fill that shell and become more stable also. When sodium and chlorine atoms are brought into the vicinity of each other, sodium quickly gives up an electron to chlorine and becomes a positive ion. Chlorine now has one more electron than it has protons so it becomes a negative chloride ion.

Electron transfer, electron sharing, and electron exchange are key ideas for remembering how molecules are formed. You find that, in nature, unlike charges attract each other so sodium and chloride ions attract each other and cling together to become one happy unit of a crystal of table salt. Sodium is no longer a neutral atom, nor is chloride. They are both happily combined ions. Stability is happiness to an atom. This is one way that atoms combine to form stable units. It is called *ionic bonding*. In ionic bonding, one or more electrons are transferred from the outer energy level of one atom to the outer energy level of another atom. If the ions pull toward each other, the resulting ionic units are more stable than the original separated atoms. Since energy is given off during the electron transfer and ion attraction process, the resulting product is more stable. This is like a ball rolling downhill. Once it is down, it is likely to stay there. It is energetically stable.

Plain English

Ionic bonding is the formation of a stable unit by the transfer of an electron from one atom to the other creating two ions that attract each other. They are more stable than they were as atoms before the transfer. The two newly formed ions attract each other, due to opposite electrical charge, thus forming a stable unit. **Covalent bonding** is the combination of two atoms to form a molecule by sharing a pair of electrons.

Other Ways of Forming Molecules

When two atoms share a pair of electrons to form a molecule, the combination is called *covalent bonding*. Atoms like hydrogen, chlorine, and oxygen exist in nature in pairs of atoms. The atoms establish a more stable particle by sharing electrons to complete an energy level. Because the first energy level is full if it contains only two electrons, two hydrogen atoms can share their electrons to complete the first energy level for each and become a more stable hydrogen molecule with symbol H_2.

Similarly, two chlorine atoms share two electrons making both atoms appear to have complete outer energy levels in that configuration as a chlorine molecule with symbol Cl_2. The more an atom's ion's electron configuration resembles the electron configuration of an inert gas, the more stable the combination becomes. Two oxygen atoms combine by making the same covalent bonding by sharing two electrons to become a more stable molecule of two atoms. You may visualize the sharing of pairs of electrons by referring to Figure 17.1. It is not intended to suggest that the atoms look like the diagram.

Newton's Figs

Hydrogen and chlorine are two elements that occur in nature as molecules. Each molecule of these elements is made up of two atoms that share two electrons in a covalent bond.

Figure 17.1

The diagram, of two hydrogen atoms and two chlorine atoms sharing a pair of electrons, aids in visualizing the formation of certain molecules.

It may help you to see that two atoms can share a pair of elections so that each atom in the combination may experience the stable configuration of an inert gas neighbor. Some of the inert gases are helium, neon, and argon. The configuration of the electrons in atoms like these is the most stable configuration that the atoms may achieve. Some molecules are made of different atoms that share electrons, but they do not share evenly. One atom might control the electron more than the other. This uneven sharing is called polar covalent bonding. In either case the sharing of electrons makes more stable electron configurations possible in the formation of a molecule by covalent bonding.

Metals Exchange Electrons

You can imagine that the atoms of solid metals are very close together. They are closely packed is a better way of stating their arrangement. They are so closely packed (each atom of the element is surrounded by 12 other atoms) that electrons are shared freely throughout the atoms.

Plain English

Metallic bonding is the name of the attraction atoms of solid metals have for each other in a closely packed arrangement due to the continuous exchange of loosely held electrons in the outer energy levels of the atoms of the metal. Metals are found on the left and bottom side of the periodic table.

The reason is that the electrons in the outer energy levels are loosely held and easily moved from one atom to the next. The sharing of electrons in this setting enables the atoms to attract each other. This type of bonding is called *metallic bonding*. You find later that the loosely held electrons that are continually exchanged are subject to being moved from one part of the solid to another by a special force. You can now identify several ways that molecules are formed. You should be aware that most of the discussion has centered around two atoms combining in some fashion to form a molecule. The same processes can involve several atoms in the formation of molecules.

Movement of Atoms and Their Components

The movement of some atoms and their components has been mentioned in this chapter, such as the nucleus of the helium atom. Alpha particles emitted by radioactive atoms travel at about 20,000 mi/s. The beta particles are high-speed electrons, meaning they travel at speeds around 80,000 mi/s and up. The vapor trails of these particles can be observed easily in a *cloud chamber*. You know that the speed of atoms and molecules of matter depends upon the state and the temperature of the material.

Plain English

A **cloud chamber** is a transparent container filled with alcohol vapor suspended between electrically charged plates. While a radioactive source emits alpha and beta particles, a source of light is directed at the alcohol-filled chamber. Trails much like vapor trails of jet planes are formed along the path where the radioactive particles have traveled. Their trails are easily viewed by the light reflected from the droplets of vapor in the cloud formed.

At S.T.P. nitrogen molecules travel at about 500 m/s. Molecules of solids vibrate about an equilibrium position with amplitudes of about .01 nanometers with frequencies of 10^{13} cycles/s. The atoms have diameters of about 0.2 nanometers and molecules range in diameter from a few nanometers to hundreds of nanometers depending on the number of atoms in the molecule. The particles of matter race around, bounce off containers, and vibrate at incredible speeds. The type of motion depends on the state of matter, and the speeds are increased by increased temperature.

It is interesting to note that not only does radioactive material emit alpha particles but also beta particles! There are no electrons in the nucleus and yet beta particles traveling at high speed and tremendous penetrating power have their origin in the nucleus.

The mechanism used to explain the emission of a beta particle is that a neutron decays into a proton that stays in the nucleus and a beta particle that flies out of the nucleus. You may want to consider the following reactions, one for alpha decay and one for beta decay.

Newton's Figs _____

The equation summarizing beta decay in this section does not include a particle that is involved in the decay of a neutron. The particle is a neutrino that, strange as it may seem, has only a small amount of mass.

$$^{226}_{88}Ra \rightarrow\ ^{222}_{86}Rn + \ ^{4}_{2}He + energy$$

$$^{60}_{27}Co \rightarrow\ ^{60}_{28}Ni + \beta$$

The alpha decay yields a daughter nucleus with a mass number that is four less than the parent nucleus and an atomic number that is two less than the parent nucleus. The beta decay yields a daughter nucleus that has the same mass number as the parent and the atomic number of the daughter nucleus is one higher than the parent nucleus. Radium has a *half-life* of about 1620 years. That means that in 1620 years a sample of 1.0 g of radium will become a mixture of 0.5 grams of unchanged radium and the other 0.5 grams of original radium sample will have decayed to other atoms.

Plain English _____

The **half-life** of a radioactive element is the time required for half of the atoms in a given sample of the element to decay.

The half-life of carbon 14 is about 5730 years, and the process of carbon dating is fairly accurate up to about 60,000 years.

The Least You Need to Know

- ◆ Interpret the meaning of atomic number and mass number.

- ◆ Describe the atom using the model outlined in this book.

- ◆ Explain the meaning of the ion and list examples of ions.

- ◆ Identify methods of formation of molecules.

- ◆ Calculate the mass of a mole of an element.

Electricity at Rest

In This Chapter

- ◆ Electrical charge at rest

- ◆ Substances that carry a charge

- ◆ Two ways to charge an electroscope

- ◆ Know when something is charged

- ◆ Danger from the clouds

We had just arrived home and ran to the front door because the threatening storm was beginning to break. Suddenly there was a bright flash of lightning and the immediate report of thunder. We knew that the strike was close. Later we learned that just a few blocks away a young man carrying aluminum baseball bats from the practice field had been hit. Even though he recovered, it took a long time for him to come to grips with the frightening experience of being hit by lightning.

You have probably heard of similar events. You will find out more about charged bodies and how they affect each other as well as your environment. You will see an electroscope charged two different ways, and you should be warned that the indication of charge is just that. There is no attempt to enable you to count the number of charges involved but you will be able to identify the effects of charging by observing the charge indicators to be introduced at the proper time in this chapter.

Static Charge

You probably have experienced a surprising snap when you walk across the floor to switch on the lights. Before pushing the elevator button, you must recover from the shock you received when you were in the process of placing an outstretched finger on the button that indicated the elevator was on the way up. Touching a keyboard or the computer on a day when sparks are snapping from your finger to a light switch is something you do not want to do. The charge you experience is unhealthy for the tiny components that are affected by the sudden surge of electricity through the computer. Electricity can be a lot of fun to explore, but you must always respect the intensity of a jolt of electricity and know that it can be lethal.

It is practically impossible to convince a reader to stop and try some of the activities of physics, but I want to try to have you do physics for a few minutes. You have probably noticed this many times in your life, but take a comb and run it through your hair. Then look at your hair in the mirror. Is some of it standing up and away from other hairs on your head? You had no success? Maybe it was a hot, humid day. This hair-raising activity works best during a dry, winter day. If the first activity didn't work, try this then, please.

Tear off a few tiny pieces of newspaper and put them on a flat surface, but not a rug please. Now, comb your hair and pass the comb near the pieces of newspaper. Did the paper fly up and stick to the comb? It did not work? Talk about a bad hair day!

Try this, I know that this will work unless you are standing in water or you are out in the rain. Tear a few strips of newspaper about two to three inches wide and fifteen to twenty inches long. Lay one of the strips on a rug, woolen shirt, or coat and stroke the strip of paper lightly with your fingers. Lift the strip of newspaper, hold it up against the wall, and let go. See that? There are no tricks and no magic. Now just to make sure, try holding another strip of newspaper against the wall that you have not stroked and let go. Now we are getting somewhere. Place two strips of the newspaper on the rug and lightly stroke them with your fingers. Pick up one of the strips in each hand and bring the flat sides close to each other. Can you believe that? It is amazing how much fun you can have with physics. It is even more fun to try to explain what has happened so that the event can be repeated with the same level of success that you have just realized.

Newton's Figs

Rubbing amber with fur creates an observable charge. You refer to this method as charging by using friction.

You have been playing with static electricity. You knew that already, I am sure. However, did you know why it is called static electricity? It is because static electricity is created by charges that are just sitting there waiting to be manipulated by some pointy-eared space alien. With your hair standing up like that, you look like you just stepped off the spaceship! The two paper strips pushed away from each other, didn't they? Pushed is the right word, I think, because as you try to move the strips closer together they move farther from each other. That is what static electricity will do for you, make your hair stand on end and curl your newspaper.

In each of these antics, you rubbed your hair with a comb or stroked the newspaper with your hand. That is a key to dealing with static electricity.

By the way, you will find that all of the business we are discussing works best on a dry day or at least not in damp weather, that is soap-bubble-blowing time. You wonder if physicists do that, blow soap bubbles? I don't know about physicists but physics teachers do, at least the one I know does. A distinguished gentleman from Oshkosh, Wisconsin, who has a Ph.D. in physics taught me how to create inverted bubbles. Inverted bubbles are thin films of air filled with water and suspended in water. Regular soap bubbles, as you know, are thin films of soap solution filled with air and suspended in air. But back to static electricity; I am often sidetracked when I get excited exploring ideas about physics with people who are interested.

Scientists have found that there are two types of electrical charge associated with static electricity. One type is associated with a glass rod and the other type is associated with a hard rubber or amber rod. Rubbing the amber rod with fur enables you to cause that rod to be charged with one type of electrical charge. The glass rod rubbed with silk allows you to charge that rod with a different kind of electrical charge. When you walk across a carpet and open a door that has a metal doorknob, you often experience a jolt. That means that walking across the carpet charged your body electrically. The jolt you experienced was either your body giving that charge to the doorknob or gaining charge from the doorknob. The term "charge" is used because the two types of electrical behaviors are associated with two types of charge. Benjamin Franklin was fascinated by electricity and conducted some unmentionable experiments that you definitely do not ever want to repeat.

CAUTION   **A Johnnie's Alert** _____

Creating or generating a charge of the size discussed thus far causes no health problems. You may experience a bit of discomfort by the discharge to the elevator button, but there is nothing life-threatening in these activities. However, there are objects like a Leyden jar or a television picture tube that can deliver a fatal jolt of charge if they are not handled properly. Please leave these critters alone unless you have had proper training in handling them.

Franklin named the two types of charge positive and negative. These are arbitrary names; that is, he could have named the charges Bill and Sue, but he chose positive and negative. These words do not have the same meaning as they do when you are talking about a number line. His theory proposed that when one type of charge is observed in one body, an equal amount of the opposite charge is observed in another body. The idea is that charge moves from one body to another, but is not lost. The algebraic symbols + and − are associated with these names to be consistent with the notion that the net charge created is zero. That is a crude statement of the conservation of charge. It does establish the important principle that ultimately no charge is lost or gained, or the total amount of charge in the universe is constant.

Insulators and Conductors

The strips of newspaper are an example of a type of matter that is called an *insulator*. Charge can be deposited on an insulator and the charges remain on the insulator where they are placed until they are wiped off for some reason. Insulators do not allow an easy movement of thermal energy. They also do not allow an easy movement of electrical charge. Remember those free-moving electrons in metals? They play a big part in conduction of both thermal energy and electrical charges. More about metals and electrical charge movement later.

Plain English

An **insulator** is a material that does not allow free movement of charge.

A Johnnie's Alert

The magnitude of the charge on the proton is the same as that on the electron. The charge on the proton is called positive and the charge on the electron is called negative. Electrons are free to move in a metal conductor while the protons remain in the relatively stationary nuclei of the atoms. Both positive and negative ions are free to move in a solution.

The amber rod, the glass rod, and the hard rubber rod are all good insulators. All of these materials are great tools for studying static charge. Once charge is deposited on the insulator, it can be transported from one place to another. That charge is easily deposited by wiping an object to be charged with the charged insulator. You know that the positive and negative charges are the same charges associated with the proton and electron, respectively. Your knowledge of the structure of the atom makes it clear why only the electrons are moved around as charge in our discussions.

The proton is fixed inside the nucleus of the atom and is not easily moved by ordinary means. That means that only the electrons are either taken off or placed on an insulator by friction with some other material like silk with glass during the charging process. When amber or hard rubber is rubbed with fur, electrons are deposited on the amber or hard rubber from the fur. That means that the amber would have more negative charges than positive charges and becomes negatively charged, leaving the fur positively charged. The fur is positively charged because it has a deficiency of electrons. The amber is negatively charged because it has excess electrons. It is interesting to note that the word electron comes from the Greek word for amber.

Conductors and Electrical Charge

Conductors are materials that allow a great deal of freedom of motion for the valence electrons. Recall the metallic bonding in which electrons are attracted to several different nuclei at the same time, thereby making them fairly mobile rather than localized as is the case in most bonding situations.

Those electrons can be pushed around the metal easily. Remember the pieces of newspaper pushing each other apart when you charged them and brought them near each other? That demonstrates what happens when two objects that have the same charge are brought near each other.

Plain English

Conductors are materials that allow electrons to move freely throughout the material.

Charging an Object

The first law of static electricity is demonstrated by those strips: Like charges repel each other, and unlike charges attract. That means that protons repel protons and attract electrons. In general, a body that has a deficiency of electrons is charged positively and an object with an excess of electrons is negatively charged. Two bodies attract each other if one is charged positively and the other has a negative charge.

Suppose that you charge a hard rubber rod negatively by rubbing it with fur. Friction helps you to deposit electrons on the insulator by rubbing them off the fur. The insulator is now negatively charged because it has an excess of electrons after being rubbed by fur. Now you may take the negatively charged insulator to the location of another insulator, rub some of the electrons off the first insulator onto the second, and then they both will be negatively charged. The second insulator has been charged by direct contact or just charged by contact. Notice that there are three characteristics of charging by contact. One is that the charged body must touch the charging body. The second is that the charged body ends up charged with the same kind of charge as the original charging body has. Finally, the body doing the charging ends up with less magnitude of charge then it had before it shared its charge with the second body. The implication is that there must be at least one other way of charging a body. There is, and you will see that method later.

Suppose that you rub the glass rod with silk so that friction enables you to remove some electrons from the rod and deposit them on the silk. The silk would be charged negatively because it would have an excess of electrons, and the glass rod would be charged positively because it would have a deficiency of electrons. What does a deficiency or an excess of electrons mean? As you know, atoms have the same number of protons as electrons. The insulators are made of atoms. Therefore, the insulator has the same number of protons as electrons in its neutral state. If you deposit electrons on a previously neutral insulator, the insulator will have more electrons than protons so the insulator has an excess of electrons and is negatively charged. If you remove electrons from a previously neutral insulator, it will have fewer electrons than protons. Therefore, the insulator will be positively charged since it has a deficiency of electrons.

A Johnnie's Alert

If an isolated conductor has just one electron in excess, the conductor is negatively charged. It would be negatively charged if it had many electrons in excess. The difference is in the magnitude of the charge, not the type of the charge.

Suppose that you rub a hard rubber rod with fur. The rubber rod insulator would be negatively charged. The excess electrons on the insulator would repel each other because like charges repel. If you take a piece of wire, a conductor, and hold it in your other hand, you could discharge the insulator by rubbing the wire across the insulator several times. The wire discharges the insulator faster if you actually drag every inch of the insulator across the wire. You essentially rake the electrons onto the conducting wire and into your body, which acts as a large reservoir, or ground.

A Johnnie's Alert

Ground or zero potential energy has the same meaning for electrons as it does for objects when considering their gravitational potential energy with respect to the surface of the earth. Electrons at ground have zero electrical potential energy just like an object lying on the surface of the earth has zero gravitational potential energy. At "ground" level, the electron has zero electrical potential energy, and the object, mass, on the surface of the earth has zero gravitational potential energy.

The wire is a good conductor so the electrons move freely from the charged insulator through the wire to your hand. Your body is a fair conductor and provides a path for the electrons to move to the earth or ground zero unless you are standing in well-insulated shoes. The earth is a humongous reservoir of electrons as well as an endless source of electrons. A conductor that provides a path for excess electrons or a deficiency of electrons discharges the charged insulator. The newspaper strips are insulators and your hand conducted electrons off to ground or from ground as you stroked the strip. You need a detector of charge to be able to tell whether the strip is positively or negatively charged.

The charges in the discussion so far are small; only a relatively few electrons are involved in charging and discharging the insulators.

Plain English

The **coulomb** is the fundamental unit of charge in the MKS system of measurement. The MKS system is the practical system for the study of electricity. A **proof-plane** is a small metal disk attached to an insulator. The disk may be rubbed on the charged insulator to sample the charge and limit the number of charges taken. The physical size of the disk limits the size of the charge to be placed on the electroscope.

Consider the results of an experiment in which it is found that 96,500 *coulombs* are carried by one mole of hydrogen ions. Remember the hydrogen ion is just one proton and is written symbolically, $_1^1 H^+$. Hydrogen ions can be moved in a solution as can sodium ions and any other ion that goes into the solution. Recall that one mole is 6.02×10^{23} of anything. That means that each hydrogen ion has

$$\frac{96,500 \, coulombs}{6.02 \times 10^{23}} = 1.6 \times 10^{-19} \text{ coulomb of charge.}$$

The charge on the proton is then $+1.6 \times 10^{-9}$ coulomb, and since the charge on the electron has the same magnitude as the charge on the proton, the charge on the electron is -1.6×10^{-19} coulomb. Since the magnitude of the charge on the electron is 1.603×10^{-19} coulomb or $1.603 \times 10^{-19} \frac{coulomb}{electron}$, then there are $\frac{1 \, electrons}{1.6 \times 10^{-19} \, coulomb} = 6.25 \times 10^{18}$ electrons. That means that the unit of charge, one coulomb, is equivalent to 6.25×10^{18} electrons. To illustrate this point, consider a bag filled with nothing but electrons. It would have 6¼ billion billion electrons. That is a whole lot of charge. Those electrons would be pushing on each other with a great deal of force, trying to get away from each other. You see now why I emphasized that the amount of charge you are playing with is small in comparison. You will use many electrons in other applications of electricity later in this book.

A Johnnie's Alert _____

Even though a small amount of charge is being transferred to the electroscope in these activities, it is best if you use a *proof-plane* to transfer a limited amount of charge to the electroscope. A heavily charged hard rubber rod could have enough charge to curl the gold leaves enough to require the replacement of one or both of the gold or foil leaves. If you are not using gold leaves or working in a humid climate, it may not be necessary for you to do this.

Detectors of Static Charge

The gold leaf electroscope or the aluminum foil electroscope is a very sensitive instrument for detecting electrical charge. It consists of a metal knob attached to a metal rod that has two leaves attached to the other end of the metal rod. The metal rod passes through an insulator that is imbedded in a metal frame with transparent glass sides. The leaves are viewed easily through the glass and the insulator isolates the metal rod with the knob at one end and the leaves at the other.

The major components of the electroscope are shown in Figure 18.1. The neutral electroscope is shown in 18.1(A) where the leaves hang limp with equal positive and negative charges. The electroscope indicates that a charge has been placed on the knob and conducted to the rest of the conducting parts, including the leaves, when the leaves spread apart.

Figure 18.1

The gold leaf electroscope is a sensitive detector of electric charge.

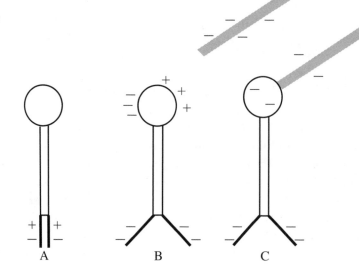

Plain English

Pith is a very light, dry, fibrous material. Pith is the central column of spongy cellular tissue in the stems and branches of some large plants. The **pith-ball** electroscope gives you an idea about the force that has been alluded to pushing the pith-balls apart. The force is very much like the gravitational force except its source is charge. The force is called the coulomb force and is calculated as $F = k\dfrac{Q_1Q_2}{r^2}$ where Q_1 and Q_2 are the charges on the two bodies, r is the distance between the charges, while k is a constant that depends on the kind of insulator between the two charged bodies. Often it is a vacuum or air.

Another good detector of static charge is the *pith-ball* electroscope. In such an electroscope, small balls of pith are suspended on insulating strings. Charge can be deposited on the pith-ball by touching it with a charged insulator. If an object of the same charge is brought near the charged pith-ball, it is repelled dramatically some distance from the charged object. Today pith-balls are not really made of plant stem tissue. Now they are made of Styrofoam, and sometimes they are spray-painted with aluminum.

If an object with an unlike charge is brought near a charged pith-ball, the pith-ball swings dramatically toward the unlike charge. That means that once charged, the pith-ball electroscope can detect like charges by moving away from a similarly charged object. There is one problem with detecting a charge with the pith-ball electroscope that you should be aware of, and you will find out more about this in the next section. If an uncharged pith-ball is brought near a charged pith-ball, the two balls will attract each other, touch, and then fly apart. Both the pith-ball electroscope and the gold leaf electroscope are good detectors of charge but the gold leaf electroscope is more versatile and requires less skill to use.

Charge by Induction and Conduction

A procedure for charging the electroscope is summarized by the diagrams in Figure 18.1B and 18.1C. The hard rubber rod with a negative charge is brought in the vicinity of the knob of the electroscope in 18.1B. The electrons on the insulator repel the electrons on the metal knob and electrons move through the conductor away from the negatively charged hard rubber rod. The electrons that shift toward the leaves of the electroscope cause the leaves to repel each other just like your newspaper strips repelled each other. That means that the leaves have like charges, and since you know that the insulator has a negative charge then the charge on the leaves is negative as long as an excess of electrons remains in the leaves. The knob on the top of the electroscope would be deficient in electrons so it would be positive until the electrons move back to their original configuration. Keep in mind that the number of protons and electrons in the electroscope has not changed. The electroscope as a whole remains neutral.

I say as long as an excess of electrons remains in the leaves because if the insulator is taken away in Figure 18.1B, the leaves will hang limp again because the electrons will shift back toward the knob rather than staying near each other in the leaves. Remember that the electrons are repulsive to each other. The negatively charged insulator is allowed to touch the knob in Figure 18.1(C), enabling electrons to actually move onto the metal knob. The electrons move to the knob because as the charged insulator approaches the knob of the electroscope, the shift in electrons discussed above occurs. The apparent positive charge on the knob that occurs as the charged insulator comes in the vicinity of the knob attracts electrons from the insulator. The electroscope is no longer neutral as a whole. It now actually has an excess of electrons. Refer to Figure 18.2 to see what happens when the insulator is taken away. Can you explain why the leaves would behave as they do?

Figure 18.2

The negatively charged electroscope results from direct contact.

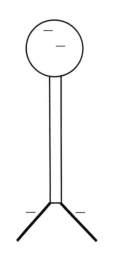

The knob collected some electrons when the insulator was in contact with it, leaving a net negative charge on the electroscope. The leaves of the electroscope spread apart indicating that they have a net charge on them.

If you bring a positively charged rod, one that has had electrons removed by friction, near the knob of the charged electroscope in 18.2, as indicated in the diagram of Figure 18.3(A), the electrons move toward one side of the knob.

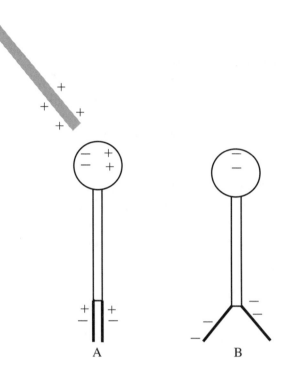

Figure 18.3

The electroscope detects charge if it is in the vicinity of the metal knob of the electroscope.

A   B

They are attracted to the positive charge on the insulator. The electrons throughout the electroscope are attracted to the positive charge that is in the vicinity of the knob. Since the leaves, connecting rod, and knob of the electroscope are all metal, the electrons move toward the area of attraction near the knob.

The leaves collapse, indicating that the net charge on the electroscope is apparently neutral. If the positively charged insulator is removed, as shown in the diagram of Figure 18.3B, the electrons flow back. The electrons move throughout the conductor, repelling each other until they are equally distributed. Then the leaves open to their original position, indicating that there is no net loss or gain of electrons due to the presence of the positive charge as long as it did not actually touch the electroscope.

The procedure outlined here is called charging the electroscope by direct contact. The charging body actually touches the electroscope.

Newton's Figs

Keep the leaves of the electroscope hanging freely so that the tiniest charge will affect the movement of the leaves. Make sure not to touch the leaves with your hands.

Charging by Induction

Another method of charging the electroscope is charging by induction. An important principle is applied in this process to accomplish the charging of the electroscope. At the proper time, a path to ground is provided for a supply of electrons. The path to ground is the body of the experimenter. Simply touch the knob of the electroscope and electrons can either flow off to ground or travel from ground onto the knob. Remember that by definition, a ground is a charge reservoir to which or from which electrons can move easily. Refer to Figure 18.4 for a step-by-step procedure of charging by induction. The electroscope is shown in Figure 18.4(A) in its neutral state. The charging body is brought in the vicinity of the knob in Figure 18.4(B), causing the leaves to spread apart because there is an excess of electrons on them. The excess electrons are the ones that were repelled down from the knob.

Figure 18.4

The first three steps in charging an electroscope by induction.

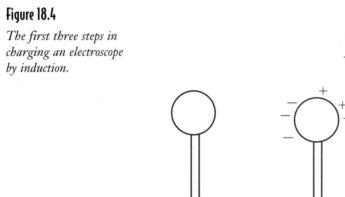

The ground connection is made in Figure 18.4C while the charging body is still in the vicinity of the knob. This means that you touch the knob of the electroscope with your finger. Excess electrons are repelled off the electroscope to ground in the direction of the arrow. They are moving away from the negative insulator. Notice that the negative insulator can affect the electrons in the electroscope even though it is not touching the electroscope. The ability to affect a body at a distance is part of the idea of force fields. We have discussed gravitational and electric force fields so far without actually giving the idea a name.

Electrons are represented symbolically as $\bar{e}$. Enough electrons are repelled to ground to make the leaves of the electroscope appear neutral because they drop down. Next the ground connection is removed from the knob in Figure 18.5(A), while the charging body is still in the vicinity of the knob and the leaves of the electroscope appear to be neutral. The leaves are spread apart in Figure 18.5(B) when the charging body is removed, showing that overall the electroscope does have a charge on it. The electrons distribute themselves evenly throughout the electroscope but there is now a deficiency of electrons. If you want to check the charge on the leaves of the electroscope, bring the negatively charged body near the knob and it will repel electrons to the leaves causing them to start to drop.

Newton's Figs

Keep the source of charge far enough away from the electroscope so that it will not leave a residual charge that is the same as the charging body because of an accidental discharge to the instrument.

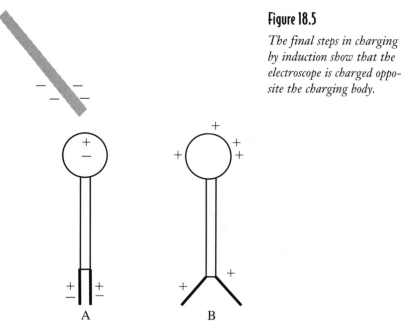

Figure 18.5

The final steps in charging by induction show that the electroscope is charged opposite the charging body.

A   B

Since some electrons were repelled off to ground and the ground was removed so that they could not return, the electroscope has fewer electrons than it had initially. The electroscope is left charged positively, the opposite charge of the charging insulator. The charging body never lost any of its original charge and never touched the electroscope during this process. This process is involved in the theory of a *capacitor*, and the same process is used to explain the topic in the next section.

Plain English _____

A **capacitor** is an electrical device used to store an electrical charge. The capacitor is made up of two conductors separated by a good insulator. The conductors are often referred to as plates of the capacitor. A **Van de Graaff generator** is a motorized source of electrons of high potential energy.

Generating Much Larger Amounts of Charge

The use of fur and an insulator to generate an electrostatic charge are fine when you are charging and discharging insulators used with the electroscope. There are times when you might want to generate a considerable amount of charge for some reason. I used a *Van de Graaff generator* for that purpose in the classroom. You may have seen demonstrations with that instrument.

The generator is essentially a motorized version of the fur and insulator method of generating electrostatic charge. The major difference is that the charge is deposited on a spherical surface at the top of the generator. After the generator runs for a minute or so there is a tremendous amount of charge on the spherical surface. You may use that large amount of charge for some spectacular demonstrations if you have the training needed to operate the generator. In order to use the generator properly, there are several things you need to know.

When a conductor (like the knob and leaves of the electroscope) is isolated by an insulator, the charge placed on the conductor distributes itself on the surface of the conductor. None of the electrostatic charge is on the inside of the conductor. It resides only on its surface. If you were to try to gather charge from the inside using a proof-plane, there would not be any to gather on the inside. Weird, isn't it? But remember the electrons are always moving so as to be as far apart as they can be. On a spherical surface, all of the charge really is evenly distributed on the surface of the sphere, but appears to act as if all of the charge is concentrated at a point at the center of the sphere.

Newton's Figs _____

The metal sphere of the generator must be clean and free of particles of any kind. Tiny particles can allow the charge to ionize the air and discharge the sphere so that it is unable to charge to capacity.

A charged spherical surface has an *electrostatic field* associated with it, as diagrammed in Figure 18.6, that is much like the gravitational field associated with the earth. One big difference is that there are two types of charge, so there are two different directions associated with electric fields. The direction is radially outward for positive charges and radially inward for negative charges.

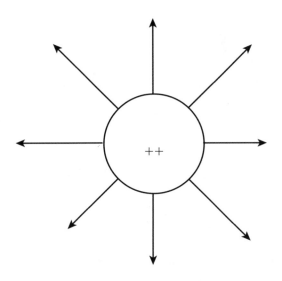

Figure 18.6

The direction of an electro-static field is shown for a positively charged conductor.

If you have a conductor with a surface that is not spherical, the electrostatic charges will repel each other in an attempt to become distributed evenly on the surface. They become very crowded on parts of the surface that tend to be pointed. If there are points on the surface, the crowding of charge becomes so intense that the air molecules become ionized. When ionization of the atmosphere near the point or points takes place, you may observe a bluish glow called a *corona* or *brush discharge* at those points. The charged object becomes discharged rapidly when ionization takes place. Sailors have observed a discharge like that emanating from sharp points on the masts of ships. They call the eerie glow St. Elmo's fire.

Plain English

A **corona discharge** or **brush discharge** is a bluish glow of ionized gases formed at any sharp point of a conductor that is under the influence of a high concentration (or density) of electrons.

Charging a Capacitor

The Van de Graaff generator provides electrons that have a high potential because of the high density of charge that you can use while it is running, or you can take electrons from the generator and store them on a capacitor. The Leyden jar is a good demonstration capacitor when it is used properly. As can be seen in Figure 18.7, the Leyden jar is a simple device, and yet it is a very dramatic piece of demonstration equipment. It is easy to use but can be very dangerous in the hands of a person who has not been properly educated to use it. The Van de Graaff generator delivers electrons that have potential energy upward to 500,000 volts.

Figure 18.7

The Leyden jar and a symbol for a capacitor are diagrammed to emphasize a method for storing charge.

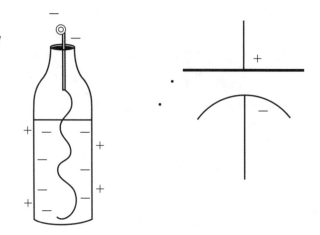

You will find later that is not too bad so long as the electrons are not delivered quickly. Once the Leyden jar is charged, most of the charge can be delivered in a fraction of a second. The same is true of the charge on a television picture tube. A lot of charge delivered in a short time is the real danger even when the voltage is low. With these cautions well understood, you may see how I used the Leyden jar in the classroom.

First, look at the symbol to the right of the Leyden jar in Figure 18.7. The symbol represents a *capacitor* in the schematic diagram of some electrical circuits.

Newton's Figs _____

Capacitors do a lot of useful things for you including helping you to tune in your favorite radio station, AM and FM.

Notice that one side of the symbol for a capacitor is positive and the curved side of the symbol is negative. Both sides are conductors and you can see that they are separated by space. The space may be filled with any kind of good insulating material. Air is a good insulator for some applications of the capacitor, such as a tuning capacitor for a radio.

The Leyden jar is a thick glass jar with the lower portion covered with a thin layer of metal both inside and outside the jar. The glass, a good insulator, separates the metal conductors. The inside conductor may be connected to the metal knob outside the top of the jar by a chain that is long enough to reach the bottom of the jar. The chain is permanently attached to one end of a metal bar that is imbedded in a wooden top for the jar. Wood is also a good insulator. The other end of the metal bar contains the metal knob where connections to the inside conductor can be made. Notice that it is a round knob and not a sharp point from which high potential electrons could escape into the atmosphere.

The metal sphere atop the Van de Graaff generator is first connected loosely to the knob or metal post of the Leyden jar. The generator is turned on and allowed to run for a short time. When the operator decides that the Leyden jar is sufficiently charged, the loose connection to the jar is broken while the generator is still running. The operator breaks the connection with a conductor mounted to an insulator at least one meter long for the sake of safety. The Leyden jar is then isolated on some sort of demonstration table where it sat while charging. It now has many electrons packed on the inside surface of the jar, charging it negatively.

The negative charge on the inside surface of the jar induces a positive charge on the outside surface. The negative charge accomplishes that by repelling electrons from the outside surface to ground. The demonstration table supporting the jar is at electrical ground potential. The atoms of the insulator become aligned so that the electrons of those atoms are largely on the side away from the negatively charged plate. That leaves the positive side of the atoms of the insulator on the side near the negative plate. The atoms of the insulator in that state of stress help to hold the negative charge on the inside plate and the positive charge on the outside plate.

After a brief explanation of the operation of a capacitor, the presenter prepares the audience for what is about to happen. She then admonishes everyone to refrain from pointing at the Leyden jar. Someone could briefly observe St. Elmo's fire at the tip of the pointy finger and then lights out for the rest of the body attached to the finger! The presenter then uses the conductor attached to the end of the meter-long insulator to approach the Leyden jar. The conductor has two spherical knobs connected through the ten-inch conductor. One knob is carefully placed in touch with the outside plate and the other knob is rotated slowly toward the knob on the Leyden jar.

Discharging a Capacitor

When the second knob of the conductor is from six to four inches away from the knob on the Leyden jar, there is a simultaneous loud crack like a large firecracker and a blue arc of the discharging electrons! The presenter inquires about everyone seeing the spectacular display of the capacitor discharging while adjusting the conductor in the same way once again. The audience observes another loud snap and a small bluish spark. She does the same thing once again with the results of a quiet snap and tiny blue spark. Then she holds the conductors in place connecting the inside and outside plates of the Leyden jar to allow them to completely discharge. She explains that the secondary sparks were displayed because the electrons rushed across the arc with so much momentum they charged the outside negatively and left the inside of the Leyden jar positively charged, and so on.

The warnings you see on the back of a television set about not opening the back unless you are a trained technician are to prevent you from discharging that giant capacitor with your body. Any time a charge is separated there is the potential that a huge arc will occur, discharging any charged bodies so they return to their neutral state. Matter is most stable when its atoms have the same number of electrons as protons. Electrons racing to neutralize a positive charge are much like water running to find its lowest level. Both water and electrons are moving toward a condition of lowest energy level. The presenter may also point out that in the presence of electrons of high potential, like those produced by the Van de Graaff generator, sharp objects in the vicinity of the generator provide a path for electrons of high potential to ionize the air and tend to discharge the generator.

One thing that is done to emphasize the effects of electrons of high potential on pointed objects is to have a volunteer stand on an insulated stool and hold the sphere on the Van de Graaff generator. I instruct the volunteer to stay on the stool until I direct them to get off. After a few moments with her hands on the generator, the hair of the volunteer usually sticks up in the air. Her new hairstyle acts as if each hair is following a line of force in the electrostatic field emanating from her head. A well-chosen volunteer with long, light hair with no hair spray helps to insure the success of this activity.

A Johnnie's Alert

If you have the opportunity to work with a Van de Graaff generator, do not allow your feet to extend over the sides of the insulated stool, because dangling toes often discharge to the ground. You should not stand too close to the table that holds the generator because discharges from your body to the table can occur.

This activity should be done only under the direction of a qualified professional. As was mentioned early in this book, activities discussed are not things you should try at home. Even under the guidance of a professional you should inform her of heart conditions or other health problems that may be aggravated by such an activity.

I then stand clear of the individual at the generator and turn the generator off. Then I hold a car key, or some other object with sharp edges, near the volunteer to discharge her before having her step down from the insulated stool. The object with sharp edges can serve useful purposes like spraying ions into the air to discharge a charged body.

Lightning and Charged Rain Clouds

Sharp objects like lightning rods can protect a house made of wood, bricks, or some other non-conducting material. The lightning rods can be placed at high strategic places on the building and, being well grounded, provide a path directly to ground for lightning. Any high object is a target for lightning. That is why I encouraged you earlier not to take cover under a tree during a thunderstorm. If you are caught out in a thunderstorm the best thing to do is get as low as possible, leaving higher targets for lightning. Any time a cloud in a rainstorm becomes charged, positively or negatively, it will induce the opposite charge on the earth.

As the intensity of the charge builds, the chance of a bolt of lightning grows. You know that a prime target for lightning is a high sharp point and even not so sharp points like your finger pointing at the Leyden jar. Lightning discharges can take place between clouds and from the earth to the cloud or from a cloud to the earth. It is said that one hundred bolts of lightning hit the earth each second. Any time charge is separated, for whatever reason, there is a good chance for a tremendous arc of energy that will occur to neutralize the charged bodies involved.

The Least You Need to Know

♦ Explain what a positively charged body means.

♦ Describe how the electroscope is used to detect a charge.

♦ Distinguish a conductor from an insulator.

♦ Identify the type and size of charge associated with the electron.

♦ Name the type of charge that is free to move in a conductor and explain why all charge is not free to move.

Chapter 19

Electricity in Motion

In This Chapter

- ◆ Moving charge
- ◆ Continuous motion of charge
- ◆ A new force
- ◆ Energy and charge
- ◆ Opposition to motion of charge
- ◆ Diverse paths for charge in motion

You considered dramatic and sometimes tragic movement of electrical charge in the last chapter. The spectacular discharging of the Leyden jar is accompanied by a loud sound and bright spark, and a bolt of lightning multiplies many times that evidence of electrical charge in motion. With knowledge of what is going on in those events, we can protect ourselves when conditions that make such an event are recognized.

We are interested in more controlled motion of charge so that we may use the energy of electrical charge in motion in our daily lives. In this chapter, applications of some methods of constraining charge in motion are discussed. You use the results of tamed electrons when you turn the lights on in your home or turn the stereo on to listen to music. Now you will take a look at just how those events are possible.

Electrons in Motion

You have just completed a review of static electricity, but I also included some electrons in motion in the last section. Any time you have electrons in motion, that means there is a current. A flow of electrons means a current; references to a flow of current do not make a lot of sense. In order for you to observe the eventual flow of electrons discharged from the inside of the Leyden jar, the Van de Graaff generator first had to do work on the electrons. That situation is very much like your lifting some object to a certain height above the surface of the earth.

Newton's Figs

Remember that work is done when a force is applied and an object is displaced in the direction of the applied force or a component of the applied force. Work done on a charge requires a force to be applied in a direction opposite to the coulomb force. That is similar to work done by the force lifting a body that acts opposite to the direction of the force of gravity on the body.

The object lifted to some height above the surface of the earth has potential energy with respect to the surface of the earth. The electrons deposited on the inside of the Leyden jar have electrical potential energy, or just electrical potential. They have potential because of the work done on each of them to force them onto the negatively charged plate of that capacitor. As the negative charge on the inside plate of the Leyden jar increased in size, the repulsive force increased, so the work done increased.

Remember, like charges repel each other with a coulomb force given by $F = k\dfrac{Q_1 Q_2}{r^2}$ where Q_1 and Q_2 are charges in coulombs, r is the distance between the charges, and k is a constant that is a property of the medium separating the charges. The units of k are $\dfrac{N-m^2}{coulomb^2}$ so that F is in newtons. The value of k for air is $9 \times 10^9 \dfrac{N-m^2}{coulomb^2}$. Notice that the force is positive if the charges are either both positive or both negative, and the force is negative if the charges are different. A positive force indicates that it is a repulsive force, and a negative force indicates a force of attraction.

You are aware already of what happens when you do work on electrons and give them a high potential. The charged Leyden jar is a good example of that situation. The electrons on the negative plate are so pumped up that if you should point at the jar, they can give you a bad hair day and curl the leaves of your electroscope! You observe both light from the arc and a frightening sound. One year my students devised a way to hold a piece of paper between the knob of the Leyden jar and the discharging conductor.

They observed the effect of a considerable amount of thermal energy as well. A charred hole made by the discharge appeared in the paper. You might find that this is not unlike dropping an object from a point above the surface of the earth. If it is released from a very high place, like a satellite, it might get quite hot or it might make a dent in the earth when it hits, like a meteorite. You would not want to be in the way of any such falling object, and that is why I cautioned you to stay out of the way of raging electrons.

Electrons in motion, electric current, can do work as they return to a lower potential. Just as you found before, potential energy can be transferred into other forms of energy like sound, heat, and light. Compare the electrons falling to lower potential to the water falling over a falls. The water can turn a turbine resulting in the production of electrical energy, as well as sound and a little thermal energy from the stored gravitational potential energy that the water had.

Current

You observed the effects of the electrons in motion, the electric current, when the Leyden jar was discharged. That discharge resulted from the ionized gas molecules providing a path for the electrons to the positive plate. In much the same way, lightning rods provide a path for the charge in a bolt of lightning to go straight to ground.

The lightning rods are conductors that contain lots of electrons that are exchanged by atoms in the conductor. The electrons are free to move from one end of the conductor to the other. The current you are concentrating on at this time is more like the electrons from the lightning rod to the ground through a metal conductor. The conductor does not have to be metal; it can be ionized air like that which enabled the discharging of the Leyden jar. But the emphasis at this time is the continuous flow of charge through metal conductors to do work on something in its path. *Electric current* is usually the quantity that is referred to that satisfies these conditions.

 A Johnnie's Alert

The spark that occurs during the discharging of charged objects like the Van de Graaff generator and the Leyden jar is amazing. It is also amazing to think that in dry air at atmospheric pressure, a difference in potential of 30 kV/cm is required to ionize the air that conducts the flow of electrons in the arc.

Physics Phun

Electric current is the flow of charge past a point in an electric circuit for each unit of time. An **electric circuit** is the conducting path for the flow of charge. The ampere is the practical unit of electric current. The **ampere** is one coulomb of charge passing a point in an electric circuit in one second. A **schematic diagram** is a symbolic representation of electric circuits. That is, symbols representing components, conductors, and instruments of measurement are used instead of pictures of those items.

Electric current is the rate of flow of charge or the flow of charge per unit of time. In symbols, $I = \dfrac{Q}{t}$, where I is current, Q is the charge in coulombs, and t is time. If the charge is one coulomb—remember that is a lot of charge, 6.25×10^{18} electrons—and the time is one second, then the current is defined to be one ampere. The ampere, $I = \dfrac{1\,C}{1\,s} = 1\,A$, is the practical unit of electric current.

The ampere is a very large unit, so more often than not you use the milliampere, mA, or the microampere, μA, as a unit of current. You know that 1 mA = 10^{-3} ampere and 1 μA = 10^{-6} ampere. Some reference books use different symbols; one popular method is 1 mA = 10^{-3}A and 1 μA = 10^{-6}A. Now that you know the units of electric current it is nice to know about instruments to measure current.

The instrument used to measure current in basic circuits is the ammeter. Other instruments used for measuring current are the milliammeter, microammeter, and the multimeter. The multimeter is an instrument that provides you with the options of measuring any one of several quantities that are found in electric circuits. The use of any of these devices to measure current requires that you connect them in the circuit so that they sample all of the current in the circuit at the place where you wish to measure it. That connection is illustrated in the *schematic diagrams* of electric circuits in Figure 19.2.

Magnetism

A conductor carrying a current of electrons has associated with it a force considered here for the first time. The new force is a *magnetic force* that results from the interaction of some object with the north pole and/or south pole of a *magnet*. The magnetic force is similar to a gravitational force or the force between electrical charges. It is similar but very different in one way. The similarities include forces that repel and/or attract. If a mass is brought near a planet, both planet and mass experience a gravitational attraction. When electric charges were discussed, it became clear that there are positive charges associated with protons. Positive charge is associated with any object that has a deficiency of electrons. Negative charge is associated with individual electrons as well as any object that has an excess of electrons. It is found that like charges repel and unlike charges attract. It was also established that a positive charge or a negative charge could be created on an isolated conductor or an insulator. Magnets have a north pole and a south pole not to be confused with positive and negative charge. The like poles repel and unlike poles attract each other. The difference in these three is that magnets exist with both poles present no matter how small the magnet may be.

Plain English

A **magnet** is an object made of iron or steel and is characterized by a north pole and a south pole each having the ability to strongly attract iron. A **magnetic force** is a force resulting from the attraction or repulsion of a magnetic pole.

Magnetic Fields

Everyone in the United States has experienced some application of a magnet at some time in his or her life. Many people have magnets sticking on the refrigerator supporting notes or pictures of loved ones. If you take one of the flexible magnets and cut it into tiny pieces, you find that the tiny pieces still exhibit the properties of the large magnet. They stick to the refrigerator! The reason being that each little piece has a north pole and a south pole that enable it to stick to the refrigerator as a result of lines of force between the two poles. You are probably more familiar with a bar magnet, a straight rectangular piece of metal containing a lot of iron. Iron acts to concentrate the magnetic lines of force and provides a path for the *magnetic lines of force.*

Plain English

The **magnetic lines of force** are imaginary lines directed from the north pole through the space near the magnet into the south pole.

The magnetic lines of force associated with an isolated magnet can be thought of as acting in a direction out of the north pole through space to its south pole. That is a common convention. It is similar to the way that lines of force associated with electrically charged bodies are thought to extend outward from a positive charge and inward toward a negative charge. You probably observed the lines of force of a bar magnet or permanent magnet early in life. If you have not yet experienced that joy, you can demonstrate the phenomenon safely in your own home. Buy a bar magnet and some iron filings from almost any novelty store or a crafts and hobby shop. Put a piece of white paper over the bar magnet that works best if placed on a wooden or glass surface at some distance from any metal. Sprinkle the iron filings over the white paper and gently shake the paper. You will see the iron filings line up in a beautiful pattern that can be imagined as a collection of lines drawn from the north pole around an oval path into the south pole of the magnet. You see the pattern in the plane of the paper but you can visualize the pattern surrounding the magnet in three dimensions. The three-dimensional pattern is evidence of the existence of the *magnetic field* of the magnet.

Plain English

The **magnetic field** is the region near the magnet containing all of the magnetic lines of force. The magnetic field is to the magnet what the gravitational field is to a planet.

Magnetic Force

The lines of force of a bar magnet can be used to describe the force on a tiny little magnetic north pole, I refer to it as a test pole, placed on one of the lines of force. The test pole experiences an attractive force from the south pole of the permanent magnet and a repulsive force by the north pole. The vector sum of those forces is the net force on the test pole due to both poles of the large magnet. The direction of the resultant force on the tiny magnetic pole is tangent to the line of force associated with the large magnet. You can think of it as the instantaneous force along the line of force in much the same way that the instantaneous velocity is tangent to the path of motion at every instant.

Since the magnetic lines are associated with a force, there must be a way to calculate the force. The magnetic force is a vector and is measured in newtons. Like the gravitational force and electrostatic force, it is an inverse square force. That is, the magnetic force is directly proportional to the product of the magnitudes of the magnetic poles and inversely proportional to the square of the distance between the poles. That means that if you place the test pole closer to the north pole than to the south pole of the permanent magnet, the force due to the north pole is larger than the force of the south pole. In general, for two like magnetic poles, the closer they are together the greater the force of repulsion; the greater the separation of like poles the smaller the value of the force of repulsion.

The Magnetic Compass

You probably know that you do not have to depend entirely on your imagination to understand lines of force for a bar magnet. More than likely you have used a *magnetic compass* to determine direction at some time in your life. The tip of the arrow of the compass points directly north and the other end of the arrow points south. The compass could be used instead of the tiny north magnetic pole to map the magnetic lines of force of a permanent magnet. The compass needle will be tangent to the line of force at each point along its path. Using the same procedure, you can map the magnetic lines of force of the earth. Since the tip of the arrow on the magnetic compass points north, the north geographic pole of the earth must be a south magnetic pole! Why?

Plain English _____

A **magnetic compass** is a device used to find direction on the surface of the earth. The tip of the needle points to geographic north and the other end points to geographic south.

The compass can help you to map the magnetic lines of force associated with the current carrying conductor mentioned at the beginning of this section. You can also get an idea about the direction of the lines of force associated with conductors by using one of two conventions. I use electron current in electrical circuits in this book but conventional current may be used to discuss electrical circuits. Conventional current has a direction opposite electron current. Both are valid conventions and you should be aware of both so that you will understand the discussions of different authors. For example, for electron current, imagine that you grasp a straight current carrying conductor with your left hand with the thumb in the direction of the current and fingers wrapped around the conductor. Your fingers point in the direction of the magnetic lines of force that accompany a current carrying conductor. If you imagine that you grasp the same conductor with the right hand with the thumb pointing in the direction of conventional current, opposite electron current, then your fingers point in the direction of the magnetic field associated with a current carrying conductor. Both methods give you the same result; the magnetic field is circular in planes perpendicular to the conductor. This result can be verified with the magnetic compass.

Magnetic Fields and Relative Motion

An interesting result is obtained if you move a straight conductor through a magnetic field perpendicular to the magnetic lines of force. Electrons move through the conductor in such a direction that a magnetic field is established around the conductor that opposes the motion of the conductor. The current in the conductor is called an induced current due to the relative motion between the conductor and magnetic field. The current in a single conductor is not very large unless the motion is high speed through a lot of lines of force. It may be easier to consider a different configuration than a single conductor. Suppose the single conductor is wrapped around a toilet paper tube to form a coil and a _galvanometer_ is connected to the ends of the conductor forming a complete circuit.

Instead of moving the conductor, imagine that you shove the north pole of a magnet into the tube. The galvanometer will indicate that a current is induced in the coil. The induced current has an associated magnetic field that opposes the motion of the north pole of the magnet. The direction of the current in the coil can be determined by wrapping the fingers of your left hand around the coil with your thumb pointing in a direction opposite the motion of the north pole of the moving magnet. The current in the coil has the direction of your fingers. The induced current has a magnetic field associated with it that causes a north pole in the direction of your thumb. The induced north pole opposes the motion of the magnet; that is, the induced north pole repels the north pole of the magnet moving into the coil. Using the right hand in the same way with the thumb pointing opposite the motion of the magnet will enable you to determine the direction of the conventional current. You find that it is just the opposite direction of the electron current.

Plain English

A **galvanometer** is a meter connected in series in an electrical circuit to indicate the presence of a current. It also indicates the direction of the current in the circuit.

A charged particle does not have to be in a conductor to interact with a magnetic field. Suppose you have two flat magnetic poles placed horizontally a few inches apart. When a stream of electrons is directed through the magnetic field perpendicular to the magnetic lines of force between the poles, you observe a change in direction of the stream of electrons. The change in direction will be in a plane perpendicular to the magnetic field. Electrons under the influence of a magnetic field with just the right speed can have the centripetal force required to travel in a perfect circle in an evacuated tube. The tube should be a type of vacuum tube so that the electrons will not lose kinetic energy by interacting with other particles.

Positively charged particles will change direction in much the same way as the electrons when they travel at high speeds in a direction perpendicular to a magnetic field. Their change in direction will be opposite the change in direction of the electrons under the same conditions. Particles like *deuterons* can be accelerated in circular paths until they have the kinetic energy desired by experimenters. Among other things, the atmosphere for the moving particles must be controlled and adjustments must be made for the increase in inertial mass of the particles as their speeds increase. Einstein predicted the increase in inertial mass of particles as their speeds increase. Particle accelerators make use of strong magnetic fields to accelerate charged particles in circular paths and are designed to provide for the increasing mass of the particles as they gain speed.

Plain English

A **deuteron** is the nucleus of an isotope of hydrogen made up of one neutron and one proton.

You must be pleased with your understanding of the ideas shared here. Did you notice all of the familiar terms and concepts involved in the discussion of these new ideas? As your interest in these topics increases, you will want to find more information about them and read more about the world of physics and especially the interaction of speeding charged particles and magnetic fields. Now you can better appreciate why it is more meaningful to talk about interactions instead of collisions. The charged particles can interact with the magnetic field and change direction without physically touching any object!

Potential Difference

The Leyden jar could be a source of the continuous current needed for an electric circuit if there was a way to keep it charged. Since charging the Leyden jar was a major task, it would be nice to have a good source of continuous current without the worry of charging a capacitor all the time.

You use several sources of continuous current in your everyday life: the C-cell, the D-cell, or the storage cells in the storage battery of your car. These cells do internal chemical work on electrons to provide them with potential energy. Think of one pole of the battery as having electrons at high potential and the other pole as being a place of low potential. You may refer to the cell as your source of continuous current with a certain potential difference associated with it. A cell also has what is called an electromotive force that includes the potential difference along with internal losses within the cell. The *potential difference* is sufficient for purposes here. The potential difference is the work done per unit charge by the chemical reaction within the cell.

Since the potential difference is the work done per unit of charge, we may find the unit's measurement for potential difference. First, the symbol for potential difference is V, and the definition for V in symbols is $V = \dfrac{W_k}{Q}$.

The units of V are joules/coulomb. If the work done is 1 joule and the charge moved is 1 coulomb, the potential difference is 1 volt, or 1 V = 1 joule/1 coulomb. Other units of measurement you might see are these: 1 millivolt (mV) = 10^{-3}V, 1 microvolt (V) = 10^{-6}V, and 1 kilovolt (kV) = 10^{3}V. Because potential difference is measured in volts, potential difference is often called the voltage. The instrument that is used to measure potential difference in a circuit is called a *voltmeter*.

Plain English

The **potential difference** between two points in an electric field is the work done per unit charge as the charge is pushed by a force in a direction opposite to the coulomb force between those points. The **voltmeter** is the instrument used to measure potential difference in an electric circuit.

A Johnnie's Alert

A small amount of current can damage the human body. Around 50 mA causes considerable pain. Around 100 mA can cause problems with the heart and even stop an unhealthy heart. Anyone who tells you that DC will not hurt you needs to know about the Leyden jar, the wrong side of a television picture tube, and a lightning bolt.

A single cell might have a potential difference of 1.5 V. You probably use cells of different sizes for your different electrical devices.

I will show you how to construct a battery by the end of this chapter. A battery may be needed to provide a potential difference of 3.0 V or 4.5 V. You may need a special battery to provide twice as much current as you get from a single cell. The current referred to here is the continuous current needed to power a flashlight or a transistor radio. It does not mean the current that exists for a very short time, which occurs if you connect the positive and negative ends of the cell together. You should never do that because not only does it ruin the cell, it can be very dangerous.

A Johnnie's Alert

The potential of the electricity from your electrical outlets in your home is about 110 V. In addition, those same outlets can supply very high continuous currents of 20 A to 30 A and higher. The continuous current in your home is AC, alternating current, which means that the electrons move first in one direction and then in the opposite direction. In the United States, the change in direction takes place every 1/120 s because a complete alternation is completed in 1/60 s. Our AC current has a frequency of 60 Hz. Direct current, DC, has only one direction in a circuit.

Resistance

Whenever an electric current is established in an electric circuit, it always experiences opposition. The *resistance* to the flow of current due to the components in an electric circuit can be thought to be like the resistance to the motion of a car due to friction.

Plain English

The **resistance** in an electric circuit is the opposition to the flow of charge or current in the circuit. A **resistor** is a component in an electrical circuit used to establish the amount of current and/or potential difference at different places in the circuit.

There is always some production of thermal energy associated with resistance in an electric circuit. This means that some of the electrical potential energy is converted to thermal energy. Sometimes sound is associated with resistance in a circuit. If the *resistor* is a filament in a lamp, there is some light in addition to much thermal energy associated with resistance.

Resistance and the Unit of Opposition

Resistance is measured in ohms, denoted by the symbol Ω. For a given potential difference, the greater the resistance in ohms the smaller the current in amperes. The resistance of a component is not measured while the component is in a circuit. Instead, at least one end of the component to be measured must be disconnected.

The instrument used for measuring resistance is the ohmmeter. Usually the resistance of a resistor is marked by a number on the resistor or indicated with a color code. Usually the color code is explained in the package of resistors. Anyway, if there is ever any doubt, you can measure the resistance with an ohmmeter.

Components of an Electric Circuit

You are now good friends with the major characters of an electric circuit: resistance, potential difference, and current. Refer to Figure 19.1 to see how all of these former strangers make up a team to play this game. The diagram in Figure 19.1 is a schematic diagram of a basic electric circuit. The break in the circuit is the symbol for a switch. No current is shown in the circuit because the switch is open or in the off position.

Always construct a circuit with a switch so that you can control the circuit. You know that current is in the circuit when the switch is closed, which makes the loop in Figure 19.1 complete. In that diagram, closing the switch causes a current in the circuit with a counterclockwise direction. The reason that is the direction is that the electrons flow from the negative side of the potential difference around the circuit to the positive side. A single cell is represented by a short and long line, labeled V in the diagram. Some disciplines use conventional positive current direction from the positive side of the potential difference around the circuit to the negative side of the potential difference.

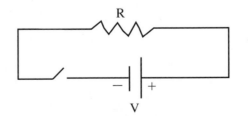

Figure 19.1

A simple electric circuit has a switch, a cell, and a load.

The current passes through a resistor, represented by a squiggly line and labeled R in the diagram, and of course through the switch when it is closed. When the switch is opened, the current stops; you have control of the switch. The symbol for current is *I*.

The current is not included in Figure 19.1 to emphasize that the switch is open. Future diagrams do indicate the direction of the current so you know that you must open the switch to stop the current.

An electric circuit has at least one resistor of some sort, usually referred to as the load, symbol R_L, or total resistance R_T. It must have a switch. It also has connecting conductors, symbolized by straight lines between components in Figure 19.1, and a source of continuous current providing a potential difference. The load resistor provides the output for the electrical energy; that is the purpose of the circuit. For example, the basic circuit can be a flashlight, so the load resistor is the light bulb that gives you the light for which you designed the circuit. The load may be made up of a string of colored lights where you can have many resistors. It can consist of several different devices provided the source of continuous current and difference of potential is sufficient. It is clear that a single cell cannot run the home refrigerator or a television set. These appliances do not use the same kind of current as is provided by a cell. A cell provides *direct current* and appliances in the United States use *alternating current*.

Newton's Figs

Use the switch in an electric circuit. Close it when you want the circuit to work and open it when you do not want current in the components. A switch can save components and measuring instruments when a connection is in error.

Plain English

Direct current, DC, is an electric current that travels in only one direction in the circuit. It travels from the negative side of the cell, through the circuit external to the cell, to the positive side of the cell. **Alternating current,** AC, is an electric current that travels in one direction and then the opposite direction with a fixed period. Your local electric company, through outlets in your home, provides AC. Although the current travels back and forth, seeming to get nowhere, the electrical energy still goes in a forward direction to your home. Otherwise, the electric company could not send you those awful electric bills.

Series Circuits

The electric circuit in Figure 19.2 is called a series circuit because the resistors are connected end to end like links in a chain.

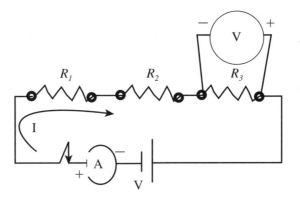

Figure 19.2

A series circuit contains resistors connected in a chain end to end so that the total current passes through each of them.

Resistors connected in a circuit in this way will all have the same amount of current pass through them. In Figure 19.2, the total current, I, in the circuit passes through each of the resistors, labeled R_1, R_2, and R_3. I connected the ammeter in series with the load, along with the voltmeter that is connected to measure the potential difference across R_3. Notice that the voltmeter is connected with its negative side closest to the negative side of the cell and its positive side closest to the positive side of the cell.

The ammeter is labeled the same way and is connected positive to positive and negative to negative. The meters are labeled in this way so that the DC current passes through the meter in the proper direction. Note well that the ammeter is not connected in the circuit in the same way as the voltmeter. The special connection of the voltmeter is discussed later.

You are familiar with the players and some of the equipment for this game of speedy electrons, all you need now are some rules of play. The rules are fairly simple but important. Therefore, pay close attention so that you can avoid trouble. First, when the electrons of the current pass through a resistor they give up potential energy. It is much like the object falling from a point above the surface of the earth; as it falls, it gives up potential energy. The electrons give up energy as they travel around the circuit. For that reason, the decrease in potential energy or decrease of potential is called a potential drop.

CAUTION

A Johnnie's Alert

The positive and negative sides of the electrical meters must be connected properly to the circuit not only so that they will measure correctly but also to protect the meter. For example, the voltmeter is wired internally so that most of the current stays in the circuit and only a small amount passes through a properly connected voltmeter.

The amount of the potential drop can be calculated using one of the rules of the game. The first rule is the relationship among resistance, current, and potential difference. The relationship is called Ohm's law. It is simply $V = IR$; the potential difference is equal to the product of the value of the resistance and the current through the resistor. The potential drop across a resistor is often called an IR drop. The second rule is stated simply as the sum of the potential drops around a circuit is equivalent to the potential drop across the cell or *battery* in the circuit. Applied to the components in the circuit of Figure 19.2, the second rule gives you $V = V_1 + V_2 + V_3 = IR_1 + IR_2 + IR_3$. This second rule is known as Kirchhoff's second law. His first law is introduced later.

Plain English

A **battery** is a combination of cells. It is constructed to overcome the limitations of one cell. A battery provides a larger current, a larger potential difference, or both. A **branch** of a circuit is a division of a parallel part of the circuit.

You may enjoy a few warm-up tosses by calculating the readings on the voltmeter and the ammeter in Figure 19.2. The total current is in each of the resistors in Figure 19.2 because they are connected end to end; the resistance of the load is the sum of the values of the resistors. The resistance of the load, or the total resistance, of any number of resistors connected in series is the sum of all of the resistors. Suppose $V = 1.5V$, $R_1 = 10.0\Omega$, $R_2 = 20.0\Omega$, $R_3 = 30.0\Omega$; calculate I, V_1, V_2, and V_3.

$$R_L = R_1 + R_2 + R_3$$

$$R_L = 10.0\Omega + 20.0\Omega + 30.0\Omega = 60.0\Omega$$

$$V = IR_L$$

$$I = \frac{V}{R_L} = \frac{V}{R_1 + R_2 + R_3} = \frac{1.5V}{60.0\Omega} = 0.025A \quad \text{By Ohm's law, } 1V = (1A)(1\Omega).$$

That means the reading on the ammeter is 0.025 A. Since $I=0.025$ A, $V_1 = IR_1 = (0.025A)(10.0\Omega) = 0.25V$, $V_2 = IR_2 = (0.025A)(20.0\Omega) = 0.50V$, and $V_3 = IR_3 = (0.025A)(30.0\Omega) = 0.75V$.

So much for the warm-up. Maybe you should check to see if you are ready to go by using Kirchhoff's second law: $V_1 + V_2 + V_3 = 0.25V + 0.50V + 0.75V = 1.5V$, and that is $V!$ It works like a charm.

Take a look at Figure 19.3, where you find a diagram of three resistors connected in parallel. Notice that the current is not the same in each of the resistors but each *branch* has its own current.

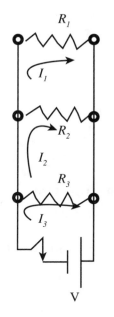

Figure 19.3

The electric circuit with three resistors connected in parallel demonstrates branch currents.

Parallel Circuits

Kirchhoff's first law is the third rule for this game, and it states that the total current passing though an area of a circuit in which resistors are connected in parallel is equal to the sum of the currents passing through the branches of the parallel part of the circuit. That is essentially a statement of conservation of charge, no charge is gained or lost when circuits branch into different paths. Kirchhoff's second law is a statement of the conservation of energy; that is, no energy is gained or lost in electric circuits. The symbolic statement of that rule for Figure 19.3 is $I = I_1 + I_2 + I_3$. Suppose the resistors and the potential difference all have the same values as in the last circuit. Calculate the total current and the branch currents in the diagram of Figure 19.3.

Notice that the potential drop across all resistors in this circuit is the same because each has one end connected to the negative side of the cell and each has the other end connected to the positive side of the cell. That means that the potential drop across each resistor is V and is stated symbolically as $V_1 = V_2 = V_3 = V$, $I = I_1 + I_2 + I_3$, $\dfrac{V}{R_L} = \dfrac{V_1}{R_1} + \dfrac{V_2}{R_2} + \dfrac{V_3}{R_3}$, $\dfrac{1}{R_L} = \dfrac{1}{R_1} + \dfrac{1}{R_2} + \dfrac{1}{R_3}$ (dividing both members by V *the potential of the cell*).

This is a general result for resistors connected in parallel. In words, it states that the reciprocal of the load or total resistance is equal to the sum of the reciprocals of each resistor connected in parallel.

The terms in the right member are added by first finding a common denominator, which is $R_1 R_2 R_3$ for this circuit: $\dfrac{1}{R_L} = \dfrac{R_2 R_3}{R_1 R_2 R_3} + \dfrac{R_1 R_3}{R_1 R_2 R_3} + \dfrac{R_1 R_2}{R_1 R_2 R_3} = \dfrac{R_1 R_2 + R_1 R_3 + R_2 R_3}{R_1 R_2 R_3}$,

$R_L = \dfrac{R_1 R_2 R_3}{R_1 R_2 + R_1 R_3 + R_2 R_3}$.

The resistance of the load in Figure 19.3 is

$R_L = \dfrac{(10.0\Omega)(20.0\Omega)(30.0\Omega)}{(10.0\Omega)(20.0\Omega) + (10.0\Omega)(30.0\Omega) + (20.0\Omega)(30.0\Omega)} = \dfrac{6000.0\Omega}{1100.0} = 5.45\Omega.$

This is another general result that you can make note of, the total resistance or equivalent resistance of resistors connected in parallel is less than the smallest resistor of the parallel resistors: $I = \dfrac{V}{R_L} = \dfrac{1.5V}{5.45\Omega} = 0.28A,\ \ I_1 = \dfrac{1.5V}{10.0\Omega} = 0.15A,\ I_2 = \dfrac{1.5V}{20.0\Omega} = 0.075A,$

and $I_3 = \dfrac{1.5V}{30.0\Omega} = 0.050A$. Use Kirchhoff's first law to check your result:

$I = 0.15A + 0.075A + 0.050A = 0.28A$. Great—that is exactly what you expected!

Series-Parallel Circuit

You have progressed from a basic circuit to a series circuit to a parallel circuit. It is time to look at the real world where you are probably always confronted by both types of connections in the same circuit. Use the same components that we started out with, except the potential difference of the cell has the same number of significant figures as the resistors—that is, $V = 150V$. Connect the resistors in a series-parallel circuit. The schematic diagram in Figure 19.4 is one arrangement.

Figure 19.4

Three resistors are connected in a series-parallel circuit.

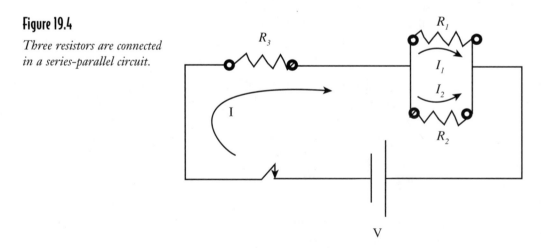

I outline the plan I use to analyze a circuit like this one. Develop your own strategy after you have practiced a bit. I begin with the most complicated part of a circuit.

In this case, the parallel resistors are combined to find an equivalent resistor. Remember that the values of the resistors are the same as before. That is, $R_1 = 10.0\Omega$, $R_2 = 20.0\Omega$, and $R_3 = 30.0\Omega$. That R_{eq} acts in the circuit like a resistor in series with R_3. I combine those resistors to find the load R_L. Using R_L and V and Ohm's law I find the current I. The IR drop across R_3 subtracted from V gives me the IR drop across the parallel resistors. Using the IR drop across the parallel resistors and Ohm's law, I find the branch currents and then the problem is solved:

$$\frac{1}{R_{eq}} = \frac{1}{R_1} + \frac{1}{R_2}$$

$$\frac{1}{R_{eq}} = \frac{R_2}{R_1 R_2} + \frac{R_1}{R_1 R_2}$$

$$R_{eq} = \frac{R_1 R_2}{R_1 + R_2}$$

$$R_{eq} = \frac{(10.0\Omega)(20.0\Omega)}{30.0\Omega} = 6.67\Omega$$

$$I = \frac{V}{R_L} = \frac{1.50v}{36.7\Omega} = 0.0409A$$

$$V_3 = IR_3 = (0.0409A)(30.0\Omega) = 1.23V$$

$$V_1 = V_2 = V - V_3 = 1.50V - 1.23V = 0.27V$$

$$I_1 = \frac{V_1}{R_1} = \frac{0.27V}{10.0\Omega} = 0.027A$$

$$I_2 = \frac{V_2}{R_2} = \frac{0.27V}{20.0\Omega} = .014A$$

That completes the solution, and now to check using Kirchhoff's first law. $I = I_1 + I_2 = 0.027A + 0.014A = 0.041A$. That is fantastic!

I promised to show you how a battery can be constructed from 1.5 V cells and now you can see how it is done. Suppose that you have a supply of 1.5 V cells that can provide 0.25 A of continuous current for a reasonable period of time. A #6 dry cell is capable of doing that, whereas the D cell or C cell or smaller are not capable of providing 0.25 A of continuous current. Before making a battery you may like to know that the positive electrode of a cell, usually the center post, is called the *anode*. The negative

Plain English

The **anode** is the positive electrode of a cell or battery and the **cathode** is the negative electrode of a cell or battery.

electrode of a cell, usually the outside container, is called the *cathode*. Using these terms can save you a lot of time when reading an explanation of how to connect cells to make batteries. You have the tools to complete the task, so we can do it.

Suppose you need a battery that supplies 3.0 V and 0.25 A. Refer to Figure 19.5 to see the 3.0 V battery needed.

Figure 19.5

A 3.0 V battery made of cells delivers the same current as a one cell.

The cell is like a resistor in that the potential differences add when connected in series. The currents add when cells or batteries are connected in parallel. In order to construct a 3.0 V battery, you simply connect two cells in series. The cells shown in Figure 19.5 are connected in series. I connected the cathode of one of the cells to the anode of the other cell but usually the connection is implied and not actually shown. Suppose you need a 3.0 V battery that supplies 1.0 A of continuous current and you are to construct it from the same supply of cells. Figure 19.6 is a schematic of the 3.0 V battery that supplies 1.0 A of current.

Notice that two cells are connected in series to have a 3.0 V battery, and then four 3.0 V batteries are connected in parallel to have 1.0 A of current since each branch contributes 0.25 A. Because the batteries are connected in parallel, the potential difference from *A* to *B* is 3.0 V. That means the anodes of all the 3.0 V batteries are connected together and all the cathodes of those batteries are connected together. The anode of the 3.0 V and 1.0 A battery is *A* and the cathode is at *B*.

You are able to construct any battery you may need from 1.5 V cells. If you require more current than one cell can supply, check the specifications of the cell or check with someone at an electronics supply store to find the continuous current rating of the cell you use.

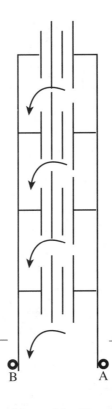

Figure 19.6

A battery that supplies current contains cells or batteries connected in parallel.

There is one other thing that you should consider if you plan on experimenting with DC. You must understand the heating effects of resistors. Resistors and other components have a power rating associated with them. Usually resistors have common ratings by physical size, such as ⅛ watt, ¼ watt, ½ watt, 1 watt, and 2 watt. The power rating insures that you use the correct component rated for the amount of current that it is supposed to handle. If the resistor has a power rating that is too small, the resistor will quickly burn out and act like a switch and open the part of the circuit in which it is connected. You have the background necessary to figure out the part played by the power of a resistor. $P = \dfrac{W_k}{t}$ and $V = \dfrac{W_k}{Q}$ (so work equals voltage times charge), $P = \dfrac{VQ}{t} = \dfrac{Q}{t}V = IV$, and $P = I(IR) = I^2 R$. You have discovered with that analysis that power can be calculated by finding the product of current and potential difference or by squaring the current and multiplying by the resistance.

Checking the power rating needed by the resistors used in the circuits constructed is an instructive exercise. All of those are checked here:

The circuit that has three resistors, 10.0Ω, 20.0Ω, and 30.0Ω, connected in series with $V = 150V$; it has a current of 0.025A. The power for R_1 is calculated as $P_1 = (0.025A)^2(10.0\Omega) = .00625W$, R_2 is calculated as $P_2 = (0.025A)^2(20.0\Omega) = 0.0125W$, R_3 is calculated as $P_3 = (0.025A)^2(30.0\Omega) = 0.0188W$. All of these resistors are safe if rated at ⅛ W or higher. Remember that ⅛ W equals 0.125 W, and each of these power values is well below that.

The circuit that has the same three resistors connected in parallel with the same potential difference has currents as listed and power as calculated: $I_1 = 0.027A$, $R_1 = 10.0\Omega$, $P_1 = (0.150A)^2(10.0\Omega) = 0.225W$, $I_2 = 0.075A$, $R_2 = 20.0\Omega$, $P_2 = (0.075A)^2(20.0\Omega) = 0.112W$, $I_3 = 0.050A$, $R_3 = 30.0\Omega$, $P_3 = (0.050A)^2(30.0\Omega) = 0.075W$. R_1 and R_2 should be at least ¼ W, and R_3 can be ⅛ W or higher.

Newton's Figs

The power rating of resistors used in a circuit or any component for that matter should be slightly larger than the calculated power to be generated in the component. The reason is that any change in current can burn out the component and open that part of the circuit.

The series parallel circuit with the same three resistors and the same potential difference has currents as listed and power as calculated: $I_1 = 0.027A$, $P_1 = (0.027A)^2(10.0\Omega) = 0.00729W$, $I_2 = 0.014A$, $P_2 = (0.014A)^2(20.0\Omega) = 0.00392W$, $I_3 = 0.041A$, $P_3 = (0.041A)^2(30.0W) = 0.054W$. All of these resistors can be ⅛ W or higher.

The Least You Need to Know

♦ Identify the necessary components of a simple electric circuit.

♦ Draw a schematic of a simple electric circuit.

♦ Calculate the current in a simple circuit given the potential difference of the cell or battery in the circuit.

♦ Explain how to calculate the load in a circuit containing resistors connected in series or connected in parallel.

♦ Use Ohm's law and Kirchhoff's laws to analyze a series-parallel circuit.

Chapter 20

Light as Particles in Motion

- ◆ Light acts like some familiar object
- ◆ Faster than anything
- ◆ Intensity of illumination and distance
- ◆ It bounces off mirrors
- ◆ It bends when it travels for air through water

Those of us who are fortunate enough to have healthy eyes are able to observe much of the beauty of the world we live in. When print begins to appear smaller and harder to read, we either have our eyes corrected with glasses or use a magnifying glass (or both). In this chapter, we will not only explain the image formed by a lens, but you will also locate that attractive image you see in the mirror when you brush your teeth. Prepare yourself by guessing where that image in the plane mirror is. Is it on the mirror, behind the mirror or in front of the mirror? Now you are ready to find out for sure.

The Particle Model

I have fond memories of the times when I would begin this unit by discussing light with my students. I told them that everybody knows how light behaves and many explain light as a particle. They were asked to share with the rest of the group and me what must be true of those particles. My class of experts always became abuzz with enthusiasm to share their ideas. This is a brief summary of the results of those sessions. The first thing noted is that if light is a particle, it must start some place such as the sun, the glow of the filament of a light bulb, or the flame of a candle. All of these sources were listed and some were eliminated from the list, such as the moon. You might pause a moment and make your own list of ideas that should be considered about these particles of light.

The question about how we perceive light always came up. That the eyes are organs that sense light in much the same way that our ears perceive sound is one idea shared by many. A question about whether it hurts or not was met by a quick negative answer because the particles are so small.

That proposal would be quickly revised by noting that sometimes it hurts when the light is too bright. There were those who would observe that the particles must travel in straight lines radially outward in all directions from a source if everyone in the room can observe the source at the same time. It was suggested that the particles sometimes got to our eyes by bouncing off objects. Furthermore, those particles that do not bounce off the object travel right past it, forming a shadow of the object. At times the particles stay on the object or sometimes travel through the object. As for those particles that travel from a source radially outward in every direction, they travel very fast—about the fastest thing that we know.

A Johnnie's Alert

Light does travel at a constant speed, but it does not travel at the same speed in all media. For example, light travels at a speed in water that is about three fourths of the speed of light in a vacuum. The speed of light in air is slightly less than the speed in a vacuum.

You can see that this particles model can generate a lot of enthusiasm in those who are interested. It must be kept in mind that it is only a model, and we would be hard put to say that light is a particle unless we agree that we are saying it behaves like a particle from our daily experience with it. Usually we reason that if light is a particle then we expect certain behaviors. We look to see if that proposal makes sense and that we are not stating, "Light absolutely is a particle." With that attitude, I approach the discussion of light with you by starting with the particle model. I explore with you several of the ideas proposed by participants in some of my former classes.

The Inverse Square Law

If light is a particle traveling in straight lines radially outward at a constant speed, you might expect to detect a certain number of particles by surrounding the source with a spherical surface.

That is the basis for experiments that have been done to see if light does travel in straight lines in all directions from the source. You need to know that there are instruments that detect light besides the eye.

A light meter is something that you may have used to see if there is enough light to take good photographs. The light meter can measure the amount of light falling on a unit area. Sources of light are rated in terms of *luminous intensity*. The *intensity of illumination* (I) is measured in *lumens*. If a light meter measures the intensity of illumination one meter from a source with a luminous intensity of one *candela*, the measurement is one lumen.

Plain English _____

The **luminous intensity** is the strength of a source of light. The unit of luminous intensity is the **candela**. The **intensity of illumination** (I) is the rate at which light energy falls on a unit area of surface. The unit of intensity of illumination is the **lumen**. One lumen is the rate at which light energy falls on one square meter of the surface of a hollow sphere having a light source at its center with luminous intensity of one candela.

Suppose you have a point source of light with particles traveling radially outward from that point as diagrammed in Figure 20.1.

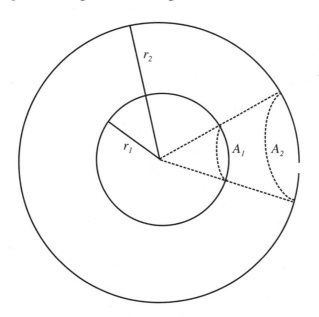

Figure 20.1

The illuminated surfaces are parts of two concentric spheres.

Imagine that we could have all the particles from the point pass through two concentric spheres. A cutaway diagram of those spheres, one with radius r_1 and the other one with radius r_2, has the point source at their center.

A cone of particles intersecting the surfaces of the spheres in areas that are directly proportional to the surface areas of the spheres A_1 and A_2 is shown with the dotted lines. Since we assume that the number of particles per unit of time remains constant at some number N as they travel outward, you can calculate the intensity of illumination of each surface as: $\dfrac{I_1}{I_2} = \dfrac{\frac{N}{A_1}}{\frac{N}{A_2}} = \dfrac{A_2}{A_1} = \dfrac{4\pi r_2^2}{4\pi r_1^2} = \dfrac{r_2^2}{r_1^2}$.

The surface area of a sphere is calculated using the formula $4\pi r^2$ and the ratio of the surface areas of the spheres is the same as the ratio of A_1 to A_2. So $\dfrac{I_1}{I_2} = \dfrac{r_2^2}{r_1^2}$ and $I_2 = \dfrac{r_1^2}{r_2^2} I_1$. If $r_2 = 2r_1$, in other words if the second surface is twice the distance from the source as the first, then $I_2 = \dfrac{r_1^2}{(2r_1)^2} I_1 = \dfrac{r_1^2}{4r_1^2} I_1 = \dfrac{I_1}{4}$. That means if you double the distance from the source, the intensity of illumination is ¼ the intensity of illumination on the first sphere. If the distance is tripled, then the intensity of illumination is ⅑ of the intensity of illumination on the first sphere. In general, the intensity of illumination, I, is inversely proportional to the square of the distance from the source. That is, $I = \dfrac{k}{r^2}$, where k is the constant of proportionality that turns out to be the luminous intensity in candela of the source of light.

Direction and Speed

The direction of the particles of light is radially outward from a point source, as you know. However, you may pick any one of those paths you wish to help you to explain a particular behavior. The paths of particular particles are called rays, and a bundle of rays is called a pencil of light or a beam of light. The particles are so small that two beams or pencils of light can cross and the particles never deviate from their paths or bounce off each other. If there were any collisions, you would see bright spots or sparks where the beams cross, but from experience you know you see only two beams.

As was mentioned earlier, the speed of these particles is the fastest speed possible as far as is known now. In fact, if a particle of matter could be accelerated to the speed of light, c, Einstein's theory predicts that the particle becomes energy according to his famous equation E = mc^2. Of course, m is the mass of the particle and c is the speed of light. The speed of light has been measured to be 3.8×10^8 m/s, accurate to about five decimal places. The speed of light in the FPS system is 186,000 mi/s.

That means that you can describe the motion of light with your knowledge of uniform motion. The distance to the sun from the earth is about 92,000,000 miles. How long does it take for light to travel from the sun to the earth? You can solve this problem the same way you solved the speed demon problem. Using that same outline, you find:

d = 92,000,000 mi

v= 186,000 mi/s

t =?

d = vt

$$t = \frac{d}{v}$$

$$t = \frac{92,000,000 mi}{186,000 mi / s}$$

$$t = 490s = 8.2 \min$$

Newton's Figs

Light is different than sound in a lot of ways, but one huge difference is that light does not need a medium for transmission.

Do you realize what that means? The sun can drop from the sky or suddenly go out, and we will not know it for a little over eight minutes! Now when you hear about those bare little particles streaking, you know that they do mean streaking through space.

At this stage, you know that the particle model can explain the direction and speed of light. It also explains that intensity of illumination varies inversely with the square of the distance from the light source. If you think about it, those particles incident on the surfaces of the spheres suggest something else. If you have particles incident on an area, there should be pressure. The particle model predicts that there should be light pressure, and as you probably know, experiments have confirmed that light does exert pressure. That gives you even more confidence in the model you are currently considering.

If a model not only explains many observations, but also predicts something that can be confirmed by more experimentation, it becomes trusted even more. As I share these ideas with you, I find myself wishing that I could go through some experiments with you. You will have to be satisfied with reading about experiments and conducting some for yourself when you can. I experiment at home all the time and experience the great feeling of elation when I get something to work!

Reflection

You can use a lot of the things discussed so far to do some armchair experimenting as you develop the ideas in this section. You have bounced a Ping-Pong ball off the table and watched the path of the ball toward the table as well as its path away from the table. If the table is nice and smooth, the path of the ball toward the table and away from the table is a straight line for the short distances it travels. The ball is a particle, and if light is a particle, it should behave in much the same way for observable distances it travels in our daily world.

Plain English

A **normal** is a line perpendicular to a line or to a surface. **Reflection** of light is the process of changing its direction when it strikes a smooth surface, causing the light to bounce off the surface.

Experiments have been done to show that whenever a light ray is incident on a reflecting surface, the light reflects in a special way and does not change speed. Before stating that, you need a brief review of some geometry. If you construct a line perpendicular to a surface, the line is said to be *normal* to the surface at the point of intersection of that line with the surface. If a plane is normal to another plane, that means the planes are perpendicular at the line of intersection of the planes.

Light is found to have two laws that it obeys when it reflects from a smooth surface:

♦ The incident ray, the reflected ray, and the normal to the surface all lie in the same plane.

♦ The angle of incidence is equal to the angle of reflection.

Reflection from a smooth surface is called regular or specular reflection. The angle of *incidence* is measured between the incident ray and the normal to the surface. The angle of reflection is the angle between the reflected ray and the normal to the surface. The normal to the surface is drawn at the point of incidence.

If the surface is not smooth, the reflection is called diffuse reflection. The light rays bounce off such surfaces at a lot of different angles because of the uneven surface. For our discussion, a plane mirror will serve as an adequately smooth surface. Refer to Figure 20.2 to see how a plane mirror forms an image.

Figure 20.2

The image formed by a plane mirror is located by reflected rays.

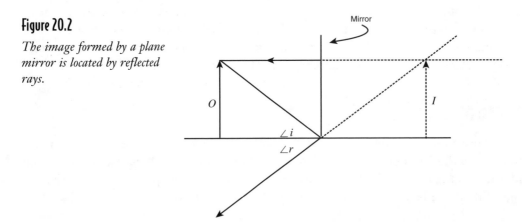

At the beginning of this chapter, I told you that you are ready to locate that beautiful image you see in the mirror. That image is located by finding the point of intersection of two virtual rays. The virtual rays appear to come from the image along the same line of sight as the reflected ray. A vertical arrow labeled *O* in the diagram represents the object, and the image is labeled *I*. We will consider only the top of the object as the source of light to make the drawing easier to understand. You may choose any ray leaving that point to help locate its image. Two rays are needed to locate the image of a point. The image point of the top of the object will be the point of intersection of the apparent or virtual rays' point of intersection. The virtual rays are the dotted lines drawn as extensions of the real reflected rays. The virtual rays extend behind the mirror where you know no rays are actually going. They can't get through the mirror, unlike Alice in Wonderland. The mirror is placed normal to the horizontal and the object is normal to the horizontal as well.

One chosen ray leaves the object incident on the mirror at an angle of incidence of 0° because you know that it reflects at the same angle coming straight back opposite the direction of incidence. The second ray is drawn from the object to the point of intersection of the mirror and the horizontal.

You know that since the horizontal is normal to the mirror, the angle of incidence is drawn equal to the angle of reflection. The reflected rays are shown with arrowheads pointed away from the mirror. The reflected rays have been extended in back of the mirror where they appear to intersect. That point of intersection is the image of the point on the object.

Newton's Figs

A real image is formed when real reflected or refracted rays of light actually intersect. A virtual image is formed when virtual extensions of reflected or refracted rays of light only appear to intersect.

All other points on the object can be located the same way, but you have probably figured out that they will be on the image so just draw the remainder of the image in perpendicular to the horizontal. The image formed by a plane, flat mirror is as far behind the mirror as the object is in front, it is a virtual image, it is right side up, and it is the same length as the object. The image is a virtual image because the light particles do not reach the point of intersection, but they appear to us when we look into the mirror to intersect behind the mirror. That description of the image is a qualitative description given by just looking at the ray diagram in Figure 20.2. A quantitative description can be calculated using geometry.

A Johnnie's Alert _____

Draw a ray diagram to locate the image formed by choosing rays of light from a point on the object that you understand. You know that a ray perpendicular to a plane mirror will reflect straight back the way it came. Draw any other ray that strikes the mirror, construct a normal at the point of incidence, and then construct the angle of reflection equal to the angle of incidence. Two rays are required to find the image of a point. Extend the reflected rays back until they intersect. That point of intersection is the image of the point source of light on the object.

Reflection and Spherical Mirrors

Light obeys the same laws of reflection any time it is incident on a smooth surface. Even if the surface is curved, the same laws enable you to locate the image formed by a curved mirror. Suppose that a mirror is cut from a portion of a hollow metal sphere in which the inside surface is polished. That mirror is called a *concave spherical mirror*. If the outside surface is polished to reflect, it is called a *convex spherical mirror*. Maybe you have looked at the outside of a shiny Christmas bulb or spherical lawn ornament. They would be examples of convex spherical mirror surfaces.

The side-view mirror on your car is a convex mirror, and the surveillance mirror in some stores is a convex mirror. The convex mirror forms only virtual images close to the surface of the mirror no matter how far away the object may be. The concave

Plain English _____

A **concave spherical mirror** is a segment of a sphere with the inside surface polished to reflect light. A **convex spherical mirror** is a segment of a sphere with the outside surface polished to reflect light.

mirror also has many applications. One is that it can act as the collector for a solar oven in needy countries where there is plenty of sun. It is probable that you have seen a concave mirror on the flip side of make-up or shaving mirrors. It is the side that magnifies. I sketched one in Figure 20.3 along with a ray diagram that locates the image formed if you have your face close to the shaving mirror. A different ray diagram is needed if you stand far away from it. Before you study the diagram you need some information about the geometry of a concave mirror.

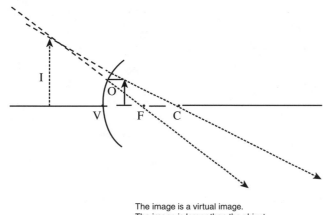

Figure 20.3

*The image formed by a con-
cave spherical mirror when
the object is between V and F.*

The image is a virtual image.
The image is larger than the object.
The image is farther from the mirror than the object.
The image is not inverted.

Newton's Figs

Your freehand drawing of ray diagrams for the concave spherical mirror should include
one ray drawn from the top point of the object parallel to the principal axis. Keep in
mind that this is a ray very near the principal axis so you know it is reflected through
the focal point. The second ray should be one that is along a line that passes through
the center of curvature of the mirror. Rays parallel to the principal axis at some distance from the principal axis reflect in front or in back of the focal point.

The mirror has a point associated with it: the center of the sphere that the mirror
came from. The point is called the *center of curvature*, *C*. The principle axis is a line
that is associated with the mirror and is one that passes through the center of curvature while intersecting the mirror perpendicularly at its center, the vertex, *V*.

Plain English

The **center of curvature** of a mirror is the center of the sphere of which the mirror is a part. The **principal axis** of a concave spherical mirror is the line that passes
through the center of curvature and intersects the mirror at a point called the vertex of
the mirror. The **focal point** of a concave spherical mirror is a point on the principal axis
where all rays intersect after they are reflected from the mirror, provided they are rays
that are parallel to the principal axis and near the principal axis. The focal point is
halfway between *V* and *C*. Therefore, the distance from *V* to *C* is 2*f*, where *f* is the
focal length of the mirror, the distance from *V* to *F*.

All of the rays of light that are close to and parallel to the principal axis are reflected back through a point called the *focal point*, F, of the mirror according to the laws of regular reflection. The object may be placed in front of the mirror in one of six regions that should be familiar to you. Those regions are:

◆ Very far away

◆ A finite distance greater than C

◆ At C

◆ Between F and C

◆ At F

◆ Between F and V

A Johnnie's Alert

A light ray that leaves the object on a line that passes through the center of curvature travels along a radius of the sphere of which the mirror is a segment. The radius of a sphere is perpendicular to the surface of the sphere at the point of intersection of the radius and the surface. That means the angle of incidence is 0°, so the angle of reflection is 0°.

The ray diagram in Figure 20.3 is for the last case. The qualitative description of the image formed is given for the object placed anywhere between V and C. Notice that two rays are used so that the image of a point, the top of the object, may be found.

Two rays are used that have predictable reflections. One ray is parallel to the principal axis and the other is a ray that passes through the center of curvature. The first ray reflects back through the focus and the second ray reflects straight back because it is traveling along a radius of the sphere, which means that the angle of incidence is 0°. The radius of a sphere is perpendicular to the surface of the sphere at the point of intersection with the sphere.

Quantitative Descriptions of Images

You can also make a quantitative description of the image, which involves actual measurements of distances such as f, S_o, and S_i. I will derive a relationship for you using Figure 20.4, and you will find that the relationship works for all cases in which an image is formed by a concave mirror. Notice that the distance from the focal point, F, to the vertex, V, is the focal length of the mirror.

The distance from F to the object is labeled S_o and the distance from the F to the image is labeled S_i. Both distances are positive if measured from F away from the mirror and negative if measured from F toward the mirror. The small error in distance represented by the line segment from V to the line representing H_o approaches zero if the rays parallel to the principal axis are very near the principal axis. Those rays are chosen so the geometry in Figure 20.4 is accurate.

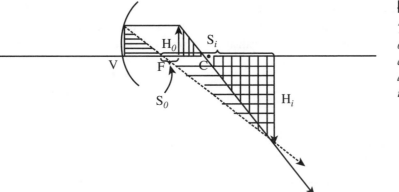

Figure 20.4

The quantitative description of the image formed by a concave spherical mirror applies to all cases when an image is formed.

The height of the object is labeled H_o and the height of the image is H_i in the diagram. The ratio $\dfrac{H_o}{H_i}$ enables you to find the magnification of the image formed by a spherical mirror.

The horizontally crosshatched triangles are similar because they are right triangles with vertical angles that are equal. That means that $\dfrac{H_o}{H_i} = \dfrac{f}{S_i}$ because corresponding sides of similar triangles are proportional.

The vertically crosshatched triangles are similar because they are right triangles with vertical angles that are equal. That means that $\dfrac{H_o}{H_i} = \dfrac{f - S_o}{S_i - f}$ because corresponding sides of similar triangles are equal.

$\dfrac{f}{S_i} = \dfrac{f - S_o}{S_i - f}$ By the transitive property of equality.

$f(S_i - f) = S_i(f - S_o)$ By multiplying both members by $S_i(S_i - f)$.

$fS_i - f^2 = fS_i - S_oS_i$ By the distributive property of multiplication.

Therefore, $S_oS_i = f^2$

Four cases of images formed by the concave mirror have been left for you to do for practice. At this time you are able to draw the ray diagram for the image formed by a plane mirror and by a concave spherical mirror. You can describe the images formed qualitatively and quantitatively.

Refraction

In the previous section light bounced back from a surface if it encountered it. There are two other possibilities: The light can be absorbed or transmitted. If light is incident on a transparent surface, some of the light is reflected according to the laws of regular reflection about which we have just talked, but part of it may be refracted. *Refraction* is our main focus at this stage of the adventure. We start with tracing the path of a ray of light as it travels from air into glass and is transmitted through a rectangular glass prism as shown in Figure 20.5.

Figure 20.5

A light ray passing from air into glass bends toward the normal.

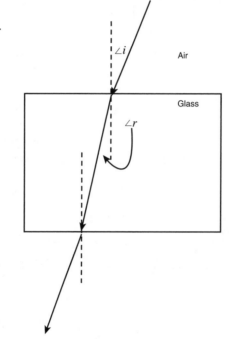

Plain English

Refraction is the change in direction of light as it leaves one medium and enters a different medium. Any time light is incident on a transparent material you will observe both refraction and reflection. You are well aware of this at night if you view the out of doors through a window. You not only see the bushes outside but also a reflection of yourself and things in the room. The reflection of the room's contents and yourself form virtual images that appear to be on the other side of the window where the bushes are really located. It is kind of weird to see a lamp sitting in your yard, isn't it?

Again we have a normal to the surface, but this time it is the surface of a glass prism. The angle between the normal and the incident ray in air is the angle of incidence and the angle between the normal and the refracted ray in glass is the angle of refraction. The angle of incidence measures about 20° and the angle of refraction is about 13°.

You can see that the angle of refraction in glass is smaller than the angle of incidence in air without measuring the angles. One way of describing this fact is that the light bends toward the normal when it leaves air and enters glass. That means that if you were to imagine extending the original ray straight into the glass, the actual ray would be closer to the normal line than the imaginary extended ray. At one time, all of the known angles of incidence and corresponding angles of refraction for different substances were recorded in volumes of books. That is much like Tycho Brahe's records of the locations of the heavenly bodies. As Kepler's laws replaced all of those volumes, the mathematical determination of Snell's law replaced all of the records of angles of incidence and corresponding angles of refraction.

Mathematics of Snell's Law

Snell's law requires that you know a small amount of mathematics beyond algebra and geometry. What you need to know is that the sine of an acute angle of a right triangle is defined to be the ratio of the side opposite the angle to the hypotenuse of the triangle. If the side opposite the acute angle is labeled a and the hypotenuse is labeled c, then

$$\text{sine } \theta = \sin\theta = \frac{a}{c} = \frac{\text{side opposite the angle}}{\text{hypotenuse}}.$$

The 90° angle is not treated the same way as the acute angles in a right triangle but there is one value that you may use right away, and that is $\sin 90° = 1$.

The simple relationship that replaced many volumes of books containing observations of angles of incidence and corresponding angles of refraction is called Snell's law and is written: $\frac{\sin \angle i}{\sin \angle r} = n$ where n is a constant, called the index of refraction, and is associated with the type of *matter* the light is entering, assuming that it is leaving a vacuum.

A Johnnie's Alert

When you use a calculator to solve problems involving Snell's law, do not be surprised if your calculator signals an error when you calculate an angle of refraction. It may be telling you that there is no angle of refraction. That means that the angle of incidence is greater than the critical angle, and you should expect total internal reflection.

The index of refraction of a vacuum is defined as 1, and for air it is almost 1. The index of refraction of glass is $\frac{3}{2}$ = 1.50 and the index of water is $\frac{4}{3}$ = 1.33. Any transparent matter has an index of refraction, but I use only these three in this book. You can look up others in reference books if you like. Diamond's high index of refraction makes it a very valuable gem.

You can check the behavior of the ray of light entering glass in Figure 20.5 by measuring the angle of incidence and using Snell's law to calculate the corresponding angle of refraction. The angle of incidence is about 20°, so using Snell's law you find that the angle of refraction is calculated by:

$$\frac{\sin \angle i}{\sin \angle r} = n_g$$

$$\sin \angle r = \frac{\sin \angle i}{n_g}$$

$$\sin \angle r = \frac{\sin 20°}{1.5}$$

$$\sin \angle r = 0.2280$$

Most calculators have sin functions on them. Have the calculator determine the sin of 20°, and then divide that number by 1.5. The answer of 0.2280 tells you that there is some angle that has a sin value of 0.2280. Find the angle whose sine is 0.2280 by using the inverse key and then sin or by using the arcsin key or the sin^{-1} key. When I calculate the value, I get about 13°.

A Johnnie's Alert

If you trace a ray of light from a medium that has an index of refraction greater than the index of refraction of the medium the ray is entering, then there will always be a critical angle. You will never have a critical angle going from a medium whose index of refraction is less than the index of refraction of the medium it is entering, such as going from air into glass.

Notice in Figure 20.5 that when the light ray leaves the bottom of the glass prism, it bends away from the normal and enters air at an angle of 20°.

That makes the ray leaving the prism parallel to the ray that entered the prism at the top. Behavior like that happens when the sides of the prism are parallel. If you apply Snell's law to the ray inside the prism at the bottom, you can calculate the size of the angle of refraction into air. You must realize that the process of a light ray leaving the prism is just the inverse of the process of a light ray entering the prism, so that Snell's law states:

$$\frac{\sin \angle i_{glass}}{\sin \angle r_{air}} = \frac{1}{n_g}$$

$$\sin \angle r_{air} = n_g \sin \angle i_{glass}$$

$$\sin \angle r_{air} = (1.5)\left(\sin 13°\right)$$

$$\sin \angle r_{air} = 0.3374$$

$$\angle r_{air} = 20°$$

The Critical Angle of Incidence

You have just used the inverse of Snell's law. However, it works only for all angles of incidence in glass such that the ray leaves the glass and enters air or a vacuum. You find that as you increase the angle of incidence in glass, the ray entering the air bends farther away from the normal. The last ray that will leave glass and enter air leaves at an angle of 90° in air. The angle in glass that would cause that angle of refraction is called the critical angle for glass. Any transparent substance has a critical angle relative to air, the last angle of incidence that has an angle of refraction in air equal to 90°. You may use Snell's law to calculate the critical angle because it is calculated the same way you did it before. That is, Snell's law states $\dfrac{\sin \angle i_{glass\,crit}}{\sin 90°} = \dfrac{1}{n_g}$ or $\sin \angle i_{glass\,crit} = \dfrac{1}{n_g}$

since the value of $\sin 90°$ is 1. That means that $\angle i_{glass\,crit} = \sin^{-1}\left(\dfrac{1}{n_g}\right)$ or $\angle i_{glass\,crit} = $ arcsin$\left(\dfrac{1}{n_g}\right)$. Both types of notation mean the critical angle of glass is equal to the inverse sine of the reciprocal of the index of refraction of glass. You can get a numerical answer using that result: $\angle i_{glass\,crit} = \sin^{-1}\left(\dfrac{1}{1.5}\right) = 41.8°$. That means that any angle of

incidence in glass less than or equal to 41.8° will leave the glass and enter air. Any angle of incidence in glass greater than 41.8° will be totally reflected back into glass. That is called total internal reflection. The ray is not able to leave the glass if the angle of incidence is greater than the critical angle. Since it stays in the glass, the laws of reflection now take over as if it bounced off a mirrored surface. Actually such reflections are even more efficient than mirrored reflections.

The last result actually applies to any transparent substance. The *critical angle* may be calculated as $\theta_{cs} = \sin^{-1}\left(\dfrac{1}{n_s}\right)$, as long as the light is entering air or a vacuum, and is read: The critical angle of the substance is equal to the arcsin of the reciprocal of the index of refraction of the substance. Calculate the critical angle for light as it tries to leave water

Plain English

The **critical angle** of a substance is the angle of incidence in the substance that has an angle of refraction of 90° in a second substance.

and enter air. $\theta_{iw} = \sin^{-1}\left(\dfrac{1}{\dfrac{4}{3}}\right) = \sin^{-1}\left(\dfrac{3}{4}\right) = 48.6°$. The last ray that will leave the surface of

water is incident at the angle of 48.6° and will enter air at an angle of refraction of 90°. Light enters water by following the inverse path. The last ray that enters water from air is at an angle of incidence of 90° and it refracts at an angle of 48.6° in water. That is why fish see the world through a cone of light defined by the critical angle. To a fish, the world must appear as a circle above their eyes! If you want a fish's eye view of the world, the next time you go swimming stand on the bottom of the pool and look up at the surface.

There is a circle through which you may see people standing at the side of the pool. Beyond that circle the surface of the water is a mirror and reflects all of the light back into the water.

Snell's Law for Two Media Other Than Air

I am sure you have noticed that I have used only substances with air to apply Snell's law. What happens if you trace a ray of light from water into glass? Even though there is no obvious answer, you can use information you have to develop an answer to the question. Suppose you have two prisms with parallel sides separated by a thin layer of air. Trace a ray of light from air through the prisms and the layer of air as shown in Figure 20.6.

Figure 20.6

A ray of light travels from medium I into medium II according to the expression $n_1 \sin\theta_1 = n_2 \sin\theta_2.$

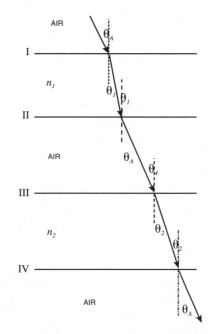

Start with the surface at I and end at surface IV.

$$\frac{\sin \theta_A}{\sin \theta_1} = n_1, \text{ at I}$$

$$\frac{\sin \theta_1}{\sin \theta_A} = \frac{1}{n_1}, \text{ at II}$$

$$\frac{\sin \theta_A}{\sin \theta_2} = n_2, \text{ at III}$$

$$\frac{\sin \theta_2}{\sin \theta_A} = \frac{1}{n_2}, \text{ at IV}$$

Both the first and second equations show that $\sin \theta_A = n_1 \sin \theta_1$ and both the third and fourth equations show that $\sin \theta_A = n_2 \sin \theta_2$. If you look at Figure 20.6, you see that all of the normals are parallel and that means all of the angles in air, θ_A, are equal.

Therefore, all of the expressions, $\sin \theta_A$, are equal. That gives you:

$$\sin \theta_A = \sin \theta_A$$

$$n_1 \sin \theta_1 = n_2 \sin \theta_2 \quad \text{By substitution.}$$

If the layer of air becomes smaller and smaller, then the medium with index of refraction n_1 is not separated from the medium with index of refraction n_2. Therefore, the expression $n_1 \sin \theta_1 = n_2 \sin \theta_2$ enables you to describe quantitatively the behavior of light when it travels from water into glass or glass into water. You may check it to see if it works for two media you are familiar with such as water and air. Tracing a ray from air into water, the expression yields: $(1) \sin \theta_A = n_w \sin \theta_w$ because $n_A = 1$, $\frac{\sin \theta_A}{\sin \theta_w} = n_w$, and that is exactly the statement of Snell's law for that situation. That means that you may use the expression $n_1 \sin \theta_1 = n_2 \sin \theta_2$ as a generalized Snell's law. It works for any transparent media.

Newton's Figs

Since the normals are all parallel because the faces of the prism are parallel, the angles in each medium are equal. The reason is that when parallel lines are cut by a transversal, the ray of light here, the alternate interior angles are equal.

We have concentrated on refraction in this section, but you know that light incident on the surface of transparent materials reflects at the surface as well as refracting into the material. You may be wondering at this time how some particles know to enter the water and some bounce off the surface. You know that the particles also change their average speed when they enter a second medium. Light travels at a constant speed but the constant average speed is different for different media. Light does not require a medium for transmission; you know it travels through space, which is mostly empty and void of all atoms, to get to the earth from the sun.

Light travels slower in glass and water than it does in air. These are ideas you need to consider as you continue to work with the particle model of light. It helps you to know how the particle model handles reflection and refraction at the same surface. Can it explain the change in speed when light travels from one medium to another? The ray diagram in Figure 20.7 shows how a single light ray *A* is partially reflected as *B* and partially refracted as *F*. The refracted portion is incident on the bottom of the glass prism and is partially reflected as *G* and partially refracted into air as *C*.

Figure 20.7

Reflection and refraction at the same surface at both parallel faces of a glass prism.

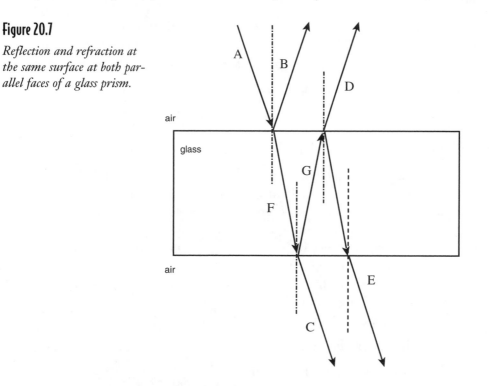

The ray *G* is incident on the upper surface of the glass prism where it is partially reflected back toward the bottom and partially refracted into air as *D*. The ray *D* emerging into air is parallel to the reflected ray *B*, and *E* emerges into air traveling parallel to *C*. Notice that part of the original ray emerges from the bottom of the glass prism traveling parallel to the original ray. It is offset a bit but leaves the bottom as if it traveled straight through the prism. If you observe this prism in the laboratory, you see that the rays emerging into air as *D*, *C*, and *E* are not nearly as bright as the original ray, *A*, incident at the top surface. It appears that no particles are lost and none are gained, and that is good. The one thing that is bothersome is that not only are these particles streaking, they must be smart little rogues as well. That may turn out to be a problem. How can they make conscious decisions?

You should trace a ray of light through at least one other shape of prism, and the glass prism in Figure 20.8 provides you the opportunity to do just that. The ray *A* incident on the prism initially is parallel to the base of the prism. The ray *B* that emerges into air is obviously deviated from the original direction of *A* to a direction toward the base of the prism. This is an interesting outcome because the final direction of the emerging ray results from refraction at two surfaces.

A Johnnie's Alert

An incident light ray that is parallel to one side of a triangular glass prism passing through the thin portion of the prism emerges into the second medium traveling more toward the thick portion of the prism.

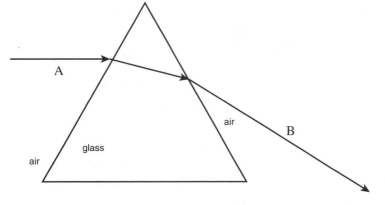

Figure 20.8

Tracing a light ray passing through a triangular prism of glass shows it emerging closer to the thicker portion of the prism.

It is instructive to note that rays through prisms with parallel faces go practically straight through the prism. Also, incident rays of light parallel to a side of a triangular prism travel through triangular prisms made of the same medium and emerge traveling toward the thicker portion of the prism. That suggests an interesting combination of prisms.

The Double Convex Lens or Converging Lens

Suppose you make a new prism with nearly parallel sides in the middle and triangular-shaped edges. You may have seen a prism like that and called it a converging or double convex lens. That is because it converges rays of parallel light into a focus area. The lens in Figure 20.9 is such a prism and uses those features to form images.

Figure 20.9

The image formed by a double convex lens can be a virtual image.

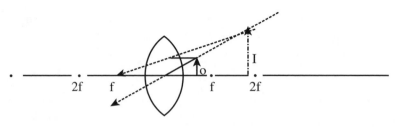

The image is between f and 2f.
The image is a virtual image.
The image is larger than the object.
The image is not inverted.

The image in the diagram is one that you have seen many times. You might not have been able to describe the image qualitatively and quantitatively with a ray diagram, but you might recognize this case as a magnifying glass. You would be looking through the lens at the object from the left side of the lens. This is like holding the magnifying lens over words in a book and looking through the lens at the words. The ray diagram shows an image formed by light that is refracted through a prism (or lens) that has the features suggested earlier. Each side of the lens is convex so the lens does not have one center of curvature like the concave mirror. Distances from the middle of the lens are measured in terms of the focal length, *f*, and twice the focal length, *2f*, where *f* is the distance to the principal focus, *F*, on both sides of the lens.

Even though you know that refraction takes place at both surfaces, the ray diagram is drawn as if all of the bending takes place at one region, the middle, of the lens. Two rays are needed to locate the image of a point. One ray is parallel to the principal axis of the lens and near the principal axis. The other ray from the object is drawn through the region of the lens where the light travels practically straight through the lens without a net change in direction. It travels nearly straight because the edges of the lens are nearly parallel there. This is like the rectangular solid you saw earlier.

A Johnnie's Alert

The double convex lens or converging lens forms both real images and virtual images. Like the concave mirror, real images are formed where real light rays actually intersect and virtual images are formed where extensions of light rays seem to converge when your brain tries to find a place of convergence. Real images are always inverted and virtual images are never inverted. The light rays that form these images are rays refracted by the lens.

The images are formed where the refracted rays intersect or appear to intersect. The image of only one point is located because images of all other points are found the same way. The image of the base of the object is located on the principal axis if the object's base was on the principal axis. The object is drawn perpendicular to the principal axis so all other points of the image are drawn with the image being perpendicular to the principal axis as well.

You have probably used this type of lens to burn your name into your baseball glove the same way my sons did when they were small. The lens can set paper on fire or inspire other forms of mischief. The reason is if you hold the lens so that the sun's rays pass through it, a tiny little image of the sun is formed at the focal point of the lens. Even though the lens is used to help correct vision like mine, the double convex lens is not the correct lens to do that because it can easily fry the retina. The moral to that remark is that you never view the sun through a lens! Use recommended ways of doing that such as viewing the reflected image of the sun formed on the end of a cardboard box with rays from the sun passing through a pinhole. The quantitative description of the images formed by a double convex or converging lens is outlined using the image in Figure 20.10.

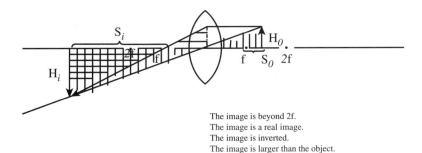

Figure 20.10

The diagram enables you to develop the quantitative description of an image formed by a double convex lens.

The image is beyond 2f.
The image is a real image.
The image is inverted.
The image is larger than the object.

Quantitative Description of Images Formed by a Lens

The distance to the image, S_i, is measured from the principal focus, on the side of the lens opposite the object, to the image. That distance is positive if it is measured from the principal focus moving away from the lens. The distance to the object, S_o, is measured from the principal focus, on the same side of the lens as the object, to the object. That distance is positive if it is measured from the principal focus moving away from the lens. All distances are measured along the *principal axis*. There are six regions where the object may be placed in order to locate the image formed with a ray diagram. The image located in Figure 20.10 makes two of the six that have been located for you.

Plain English

The **principal axis** of a lens is the line passing through the principal focus and the center of the lens. The expression derived to describe the images formed by the converging lens, $S_o S_i = f^2$, is the same expression you found for describing the images formed by a concave mirror. The principal focus is the reference point for measuring distances. The lens has two principal focal points because it makes no difference which side of the lens is used. The distances for the object are measured from the principal focus on the same side of the lens as the object. The other principal focus is the reference point for measuring distances for the image.

You may have fun drawing the ray diagrams and describing the other four images qualitatively and quantitatively. Please follow the development of the quantitative description of the image formed in Figure 20.10.

The right triangles that are marked with vertical lines are similar. The reason is an acute angle of one is equal to an acute angle of the other since vertical angles are equal when two lines intersect.

$$\frac{H_o}{H_i} = \frac{f + S_o}{f + S_i}$$ Corresponding sides of similar triangles are proportional.

The two right triangles marked with horizontal lines are similar. The reason is an acute angle of one is equal to an acute angle of the other since the vertical angles are equal when two lines intersect.

$$\frac{H_o}{H_i} = \frac{f}{S_i}$$ Corresponding sides of similar triangles are proportional.

$$\frac{f}{S_i} = \frac{f + S_o}{f + S_i}$$ Transitive property of equality.

$$f(f + S_i) = S_i(f + S_o)$$ Multiplying both members by $S_i(f + S_i)$.

$$f^2 + fS_i = fS_i + S_o S_i$$ Distributive property of multiplication.

$$f^2 = S_o S_i$$ Subtract fS_i from each member.

All the information you need is two of the three variables in that expression, and then you can find the third. In addition, you have four algebraic expressions from similar triangles that can help you to find some of the information you need.

Testing the Particle Model

How can particles know to change directions when traveling from one medium to another? My students did an experiment at this stage of development as many other students do today. Since the problem seems to be refraction, an experiment can give the student an answer. The experiment involves treating light as a particle that is familiar and checking to see if Snell's law is obeyed by the particle. The particle is a large steel ball.

The media are modeled by two horizontal surfaces separated by an inclined plane about a textbook thickness high. The particle is launched at the same speed on the upper surface and is allowed to roll down the inclined plane to the lower surface. The path of the ball on both surfaces is recorded by carbon paper tracks on plain white paper.

A Johnnie's Alert

While the particle model does not explain refraction completely, it does emphasize the importance of the change in speed of light when it travels from one medium to another.

The ball is rolled over the top of the carbon paper several times at several different angles of incidence. The angles of incidence are matched up with the corresponding angles of refraction. Snell's law does explain the change in direction of the ball on the lower surface compared to the upper surface. The angles on the lower surface were smaller than the angles on the upper surface, so the upper surface is a model of air and the lower surface water or glass.

Snell's law does explain the change in direction. You know that there is a problem here even though the change in direction can be explained. You know that if the ball rolls down the inclined plane it travels faster on the lower surface than it does on the upper surface. That just does not agree with your experience because you know that light travels faster in air than it does in water.

The particle model for light does not do a complete job of explaining the behavior of light. The first inclination is to toss it out because it cannot explain refraction. But look at all of the things it does help you to explain. The logical thing to do is continue using the particle model to explain things that it helps you to explain. Then look for a better model that can handle everything the particle model makes clear as well as things the particle model cannot explain. You do exactly that in the next chapter. You begin the search for a more complete model for light.

The Least You Need to Know

◆ Use a ray diagram to locate the image of an object in a plane mirror.

◆ Describe an image formed by a concave spherical mirror both qualitatively and quantitatively.

◆ Draw the ray diagram and describe the image of an object formed by a double convex lens.

◆ Identify the form of Snell's law needed to trace a ray of light from water into glass.

◆ Discuss any shortcomings the particle model of light may have.

Chapter 21

Light as Waves

- ◆ Waves bounce also
- ◆ Waves change direction
- ◆ Waves spread out when passing through small openings
- ◆ Waves build up and tear down

You studied longitudinal waves as a model for explaining properties of sound. At the same time you looked at transverse waves that are characterized by crests and troughs. You are ready to consider water waves, which you already know have crests and troughs. You may not know that only the wave travels through the water while the water particles move only in tiny circles.

The wave is now considered as a possible model for light. You will look at waves in springs and strings as well as waves in water in this chapter. All of these activities are meant to provide you with information so that you can decide whether the wave model is a possible model for explaining properties of light.

Reflection

You know a lot about reflection after the last chapter, but you were using the particle model of light to find images formed by a lens and mirrors.

That work led you to the notion that you should look for a better model for light. The model considered here is the wave model. You know quite a bit about waves after your work with sound. You found that there are different kinds of waves, including transverse waves and longitudinal waves. We will consider transverse waves again to review the general properties of waves. If you want to use the wave as a model for light, you need as much information about waves as you can get.

You had so much experience with particles that you realized that if light is a particle, then there must be light pressure. I told you that experiments have been done to show that light does exert pressure.

You can experience more pleasant surprises like that with waves if you gain a lot of experience with the idea of a wave. My desire is to share some ideas and experiences with you that will make the wave as familiar to you as a baseball is to your favorite pitcher.

Like many teachers of physics around the country, I gave several groups of my students two springs and some strong string to use to study waves. I showed them how to lay the spring on the floor and shake a transverse pulse into the spring for observation. After a little practice, they were on their way and found out several interesting things about waves. A pulse keeps its shape as it travels along the spring at a constant speed. The speed does not depend on the size of the pulse but it does depend on the tension in the spring. The more tension in the spring the greater the speed. They found that two different springs with the same tension have pulses traveling at different speeds. One of the springs had a small diameter and was quite heavy. The speed of the pulse in the small diameter spring was less than the speed in the spring with the larger diameter. Pulses were launched from both ends and the pulses passed through each other remaining unchanged. That is a lot like two beams of light passing through each other with no change in the beams. When two pulses from both ends of the spring meet, they combine to make a larger pulse at that instant if they are on the same side of the spring. If they are on opposite sides of the spring, they combine and produce a smaller pulse at the instant they meet.

If you think about it, the combining of pulses makes sense because the pulses displace the particles in the spring. Displacements are vector quantities, so when the pulses are on the same side of the spring, the combination would be a large displacement. Similarly, pulses meeting on opposite sides of the spring would yield a smaller displacement. The students were excited when they held one end of the spring tightly anchored to the floor and sent a pulse from the other end. When the pulse arrived at the anchored end, it reflected! It bounced off the anchored end; not only that but it reflected on the opposite side of the spring. The same thing happened for both the small diameter spring and the large diameter spring. The two pulses shown in Figure 21.1 traveling toward the anchored end in *IA* were launched rapidly, one after the other.

The first pulse reflected off the anchored end in *IB* and the second pulse is approaching that same junction. The reflection is very much like the reflection of light with the added information that the reflection is inverted when the *junction* is an anchored end.

Plain English

A **junction** is where two media are joined together. That is, a spring can be anchored at a junction, two springs can be attached at a junction, or a spring can be attached to the string at a junction.

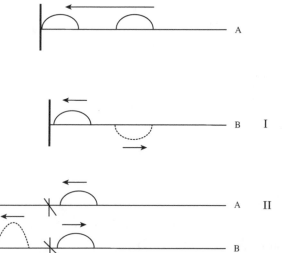

Figure 21.1

The waves reflected at different junctions are reflected differently.

Behavior of a Wave at a Junction of Two Media

A surprising thing happened when a spring was tied to several meters of string. Figure 21.1 shows a pulse in *IIA* traveling toward the junction that is marked with an *X*. The pulse is partially reflected from the junction and partially transmitted across the junction as shown in *IIB*. In addition, the reflected pulse is on the same side of the spring as the original pulse. The part of the original pulse that was transmitted across the junction was quite large and traveled very fast!

The string was tested by itself and it was found that a pulse travels several times as fast in the string as in either spring. The two springs were connected to check to see if the

A Johnnie's Alert

When two springs are connected together or a string and a spring are connected, the tension in both media is the same when tension is applied to either, provided the other end is held down.

same thing happened as with the string. A pulse traveling from the large diameter spring to the smaller diameter spring reflected at the junction on the opposite side of the spring and part of the pulse was transmitted into the smaller diameter spring at a slower speed. That is consistent with our earlier observation that the speed of the pulse is slower in the smaller diameter spring than in the larger one when they are checked separately. When a pulse was sent from the small diameter spring into the larger diameter spring, the pulse reflected on the same side as the original pulse. Part of the pulse was transmitted across the junction traveling at a faster speed. The observations are very much like partial reflection and partial refraction, but you will need a two dimensional wave to check for a change in direction.

Plain English

The **propagation** of a wave is the sending out or spreading out of the wave from a source. Waves are said to **superpose** when one adds onto the other. During the instant of superposition, neither pulse is recognizable individually because they appear as one combination of pulses.

You know quite a bit about waves in one dimension after the experience with waves in springs. You know that in a given medium the speed of *propagation* of the wave is constant. When waves pass through each other, they *superpose*, forming a pulse that is composed of two pulses while they are in the process of moving through each other. Pulses definitely reflect when they reach a junction between different media.

Transverse Waves That Superpose

Part of the pulse is not reflected but instead is transmitted across the junction. Before you can recognize that behavior as refraction, you need to observe whether or not there is a change in direction. That may happen when you look at a two-dimensional wave. Before approaching that, look at a chain of periodic waves like the ones discussed when sound was the emphasized except that sound is a longitudinal wave. The waves discussed here are transverse with a frequency and wavelength established by the source generating the waves. Keep in mind that the series of waves is traveling at a constant speed and you are able to consider only some of the waves. In Figure 21.2, there is a sequence of diagrams where the waves move a quarter of a wavelength at a time for the purpose of discussion.

In Figure 21.2(A), two complete periodic waves are shown incident on a junction where the end of the spring is anchored to the floor. The medium containing the chain of waves is shown as a horizontal straight line. You know that the particles of the medium are vibrating perpendicular to the direction of propagation of the wave. The wave has traveled to the left a quarter of a wavelength as indicated by the solid arrow above the chain in Figure 21.2(B). At that instant, the part that is reflected, a quarter of a wavelength, is indicated by the dotted arrow below the reflection. The wave superposes with its reflection, so at the instant shown the medium is horizontal because the sum of the incident pulse and its reflection is zero. The chain has moved a half of a wavelength in Figure 21.2(C), and the reflection is half of a wavelength. The reflection and its reflection are identical but the reflection is shown slightly removed from the incident pulse so that you can see it. The two pulses superpose and just at that instant you see neither pulse but rather their combination, indicated as a larger dotted pulse. Keep in mind that the train of pulses is moving at a constant speed and so is the reflection. I give you a snapshot every quarter wavelength so that you can get an idea what is happening.

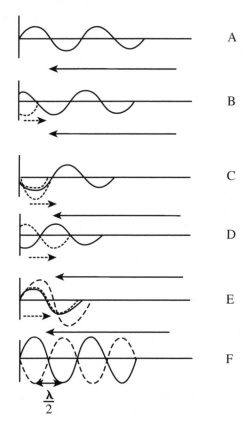

Figure 21.2

The waves reflected from a fixed junction superpose to form a standing wave.

The incident chain of waves continues to move until three fourths of a wavelength is reflected, as shown in Figure 21.2(D). At that instant, the medium would be horizontal in that region because the pulses superpose yielding a sum of zero displacement. In Figure 21.2(E), a full wavelength is reflected and superposes with a full wavelength of the incident wave.

You see neither pulse because the waves superpose, giving a sum that is a larger pulse in both regions of the incident wave. The chain of pulses continues to move to the junction superposing with a chain of pulses reflected from the junction until the pattern, shown in Figure 21.2(F), is seen. You see neither the incident chain nor the reflected chain. Instead, if the medium has the proper length, you see the superposition of the two chains of pulses. The pattern will appear to flash first the dark part of the pattern then immediately the dotted pattern. Regions called *loops* or *anti-nodes* that are a wavelength apart as shown, separated by *nodes* or *nodal points*, characterize the total pattern. The pattern is called a *standing wave*. The wavelength of the standing wave is one half the wavelength of the source. You observed the results of a standing wave in sound, a longitudinal wave, when conditions for resonance were established. The standing wave of sound is observed as a reinforced or louder sound of the same frequency as the source.

A Johnnie's Alert

The waves reflected from the barrier appear to be coming from a virtual image of the source on the other side of the junction. The reflected waves have the same frequency as the source.

Plain English

A **standing wave** is the superposition of two waves of the same frequency moving in different directions in the same medium. A standing wave can be established in a medium of proper length. Resonance of sound in a closed pipe occurs for lengths of a quarter wavelength or odd integral multiples of quarter wavelengths. A **loop** or **anti-node** is the region of reinforced amplitude in a standing wave pattern. A **node** or **nodal point** is the point in a standing wave pattern where the superpositions of the waves produce zero amplitude. Nodes are half a wavelength of the source apart, as are the loops of a standing wave.

It appears that the incident waves meet waves from a virtual source beyond the junction that has the same frequency as the source. The waves from the virtual source meet the waves from the source at the junction in such a way that a node is formed at the junction between the two. The standing wave that results from the superposition can be measured by measuring the distance between two adjacent loops. Since the wavelength of the standing wave is half the wavelength of the source, doubling that measurement determines the wavelength of the source and the wavelength of the virtual source.

Calculating Speed from a Standing Wave Pattern

If you determine the wavelength of a standing wave and the frequency of the source, you may calculate the speed of the wave in the medium using the relationship we developed earlier for periodic waves, $v = f\lambda$. Recall that I suggested you use a tuning fork held near a reflecting surface and you or a fellow experimenter can observe regions of reinforced sound and regions of practically no sound? Notice the similarity of that experiment and the standing wave. The major difference is that the standing wave is established along a straight line and sound is observed in three dimensions. The two phenomena are very similar as you find when you consider three-dimensional space.

You have noticed that I depend on a vivid imagination on your part. So before we jump to three dimensions, I act as your guide in a transition from one dimension to two dimensions. Suppose that you take a straight line and move it along a straight path perpendicular to the line. Do you see that you generate a plane? Good! Now you may put that vivid imagination to work again. Suppose that you take a standing wave pattern and move it along a straight line perpendicular to the medium of the standing wavy pattern.

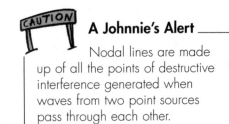

A Johnnie's Alert

Nodal lines are made up of all the points of destructive interference generated when waves from two point sources pass through each other.

Do you imagine higher portions of a plane separated by lines of lower portions of the plane? Those lines are called nodal lines. They are lines made up of all the nodes or nodal points in the standing wave pattern. The higher portions of the generated surface are made up of all of the loops or anti-nodes of the standing wave pattern. If all of this is clear to you, fine; you have made the transition from one dimension to two dimensions. Let these ideas rattle around for a while as I share a few more experiences with you.

Waves in Two Dimensions

You know that you need a two-dimensional wave to check a change in direction for refraction, if there is a change in direction. A good way to begin consideration of a two-dimensional wave is to use a ripple tank. A *ripple tank* is a piece of laboratory equipment used to study two-dimensional waves.

Plain English

A **ripple tank** is a laboratory instrument used to study two-dimensional waves. It is a shallow tank that holds water about one to two centimeters deep. It has a glass or clear plastic bottom about 24 inches square.

It is a shallow tank that holds water about one or two centimeters deep. Waves are created in the water. A light mounted above the tank casts shadows of parts of water waves on a screen placed about 20 centimeters or so below the tank. Parts of waves in the tank are higher than other regions of the water surface. The higher parts of waves act like converging lenses and bright lines are visible on the screen where the higher regions or crests of water waves focus the light on the screen.

You know that a two-dimensional wave is needed to continue your search, so two types of waves are created. You have observed the waves created by a stone hitting the surface of water so the first wave generated is a familiar one. When the tip of the finger is dipped in the ripple tank, a nice, neat circular wave travels from your finger across the surface.

Newton's Figs _____

The shape of the wave reveals the fact that the speed of a water wave is constant in a given medium. The straight wave remains straight and the circular wave remains circular until interaction with a barrier occurs.

Your finger serves as the source of waves that travel radially outward at a constant speed in all directions in a plane. You know that the speed is constant because the wave remains a circle until it strikes the sides of the tank. A plastic ruler or a wooden dowel may be used to touch the surface of the water to generate a straight wave. A straight wave travels across the tank at a constant speed. Again, you know that the speed is constant because the wave remains straight until it strikes the sides of the tank.

Reflection of Two-Dimensional Waves

A barrier placed in the tank serves the same purpose as anchoring the spring to the floor. The straight waves striking the barrier reflect as shown in Figure 21.3.

Figure 21.3

Straight waves reflect from a barrier in a ripple tank.

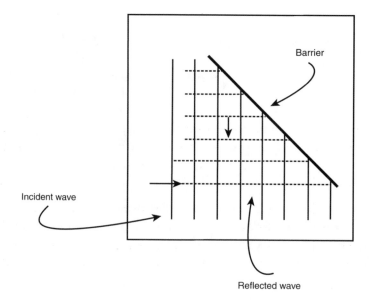

The reflection of the straight waves is a little different than the reflection of rays of light, but remember this is a two-dimensional wave. Dark lines in Figure 21.3 represent the crests of the incident waves. The corresponding crests of reflected waves are represented by dotted lines. Arrows perpendicular to a crest of each indicates the direction of propagation of each set of waves. A closer look shows that the angle of incidence is about equal to the angle of reflection. These angles are measured differently than the angles for rays. The angle of incidence is the smaller angle between the incident wave and the barrier.

The angle of reflection is the smaller angle between the reflected wave and the barrier. If you trace the path of a small segment of each wave, the paths behave very much like rays of light. The straight waves reflecting from the barrier are very similar to the light rays reflected from a plane mirror.

Waves Reflected from a Curved Mirror

A plane mirror may be easy to model in the ripple tank but a model of a curved mirror is a little more challenging. Either a piece of flexible metal or a piece of rubber tubing can be placed in the ripple tank to approximate reflection from a curved mirror. The curved mirror in Figure 21.4 yields a surprising result.

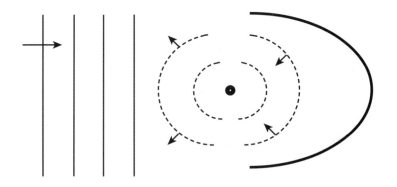

Figure 21.4

Straight waves reflect from a barrier shaped as a curve.

Straight waves traveling toward the curved barrier, as shown, are reflected as circular waves that come to focus at a point. The reflected waves pass through that point and leave as an ever-widening circular wave. If you drew the case for an object placed very far away from the spherical mirror, you found that the image is a point at the focus.

The rays locating the image intersect at the focus and continue on in a straight line radially outward from the focal point. That case is very much like the straight waves reflected from the curved mirror. Trace the path of a small segment of the straight wave to the mirror and its reflection from the mirror. The path of the segment looks very much like a ray parallel to the principal axis reflected through the focus of a spherical mirror. All of the information gathered so far about waves in two dimensions shows promise of explaining light concerning reflection, speed, and direction of propagation. You still need to find out about refraction.

Refraction

The depth of water in the ripple tank can be changed so that there are extreme differences in depth by placing a glass or plastic plate just under the surface of the water. That was done for the diagram in Figure 21.5. The dark lines represent crests of straight waves moving from left to right in the diagram as indicated.

Figure 21.5

When straight waves change media, from deep to shallow water, the wavelength changes.

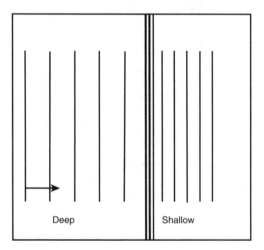

The line across the tank represents the change in depth of water from deep to shallow, reading left to right. The waves are periodic waves so both the incident wave in the deep water and the waves in the shallow water have the same frequency. Different depths of water are different media. You can see that the wavelength in the deep water is greater than the wavelength in the shallow water, so $\lambda_d > \lambda_s$, $f\lambda_d > f\lambda_s$, and $v_d > v_s$.

Since the frequencies are the same and the wavelength in the deep water is greater than the wavelength in the shallow water, the speed is greater in the deep water. The speed is frequency times wavelength. That is similar to the pulse in the large diameter spring being partially transmitted into the small diameter spring where its speed was less. That is great but you still do not know whether there is a change in direction or not.

The waves in Figure 21.6 cross a barrier at an angle. The water is deep on one side of the barrier and shallow on the other. The diagram in Figure 21.6 shows clearly that there is a change in direction as well as a change is speed. Changing the angle of the barrier for several trials reveals that Snell's law describes the change in direction and more. Look at the blown-up version of one incident wave and its corresponding refracted wave in the lower left portion of Figure 21.6.

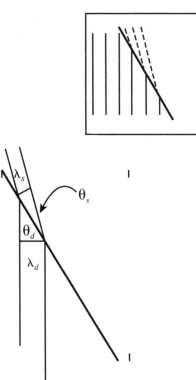

Figure 21.6

Straight waves change direction when they cross a barrier at an angle traveling from deep water to shallow.

The angles of refraction and incidence are measured the same way that the angles of reflection and incidence are measured. The angles to be measured are the smaller of the two angles between the wave and the barrier. I have drawn a line segment representing the wavelength of each wave along with a label. Two right triangles are formed; when that is done, and the triangles share one hypotenuse, call it *H*. An expression for Snell's law may be written using the two right triangles:

A Johnnie's Alert

The angle of incidence and the angle of refraction for waves is measured in a special way. The smaller of the two angles between the incident wave and the junction is the angle measured as the angle of incidence. The smaller of the two angles between the refracted wave and the junction is the angle measured as the angle of refraction.

$$\frac{\sin\theta_d}{\sin\theta_s} = \frac{\dfrac{\lambda_d}{H}}{\dfrac{\lambda_s}{H}}$$ By the definition of the sine of an angle.

$$\frac{\sin\theta_d}{\sin\theta_s} = \frac{\lambda_d}{\lambda_s}$$ Simplifying common factors.

$$\frac{\sin\theta_d}{\sin\theta_s} = \frac{\lambda_d f}{\lambda_s f}$$ Multiplying numerator and denominator by f.

$$\frac{\sin\theta_d}{\sin\theta_s} = \frac{v_d}{v_s}$$

The wave model predicts that since the ratio of the sine of the angle of incidence to the sine of the angle of refraction is greater than one, that implies the ratio of speeds is greater than one.

Newton's Figs

The wave model correctly predicts the change in speed when the wave changes media. The change in direction is correctly explained with Snell's law.

That means that the speed in the second medium is less than the speed in the first medium when the angle in the second medium is less than the angle in the first medium. That is just the opposite of the prediction of the particle model. The wave model shows a lot of promise because waves crossing a junction between media change direction and have a correct change in speed.

Diffraction

Not only do waves change direction and speed when they change media but they also bend around corners. That is called *diffraction*. Straight waves are generated in the ripple tank as shown in Figure 21.7. A barrier with a slit smaller than the wavelength of the wave is placed in the path of the straight waves.

Plain English

Diffraction is the curving of waves as they pass through an opening or pass by a sharp corner. The diffraction of light is an important property of waves. Among other things, it provides the key to checking to see if you can measure the wavelength of light.

The waves pass through the opening and curve, bending around the edges of the slit. The slit acts very much like a point source of waves. Remember when dipping the finger into the water created waves on the surface of the water?

The slit acts very much the same way if it is small compared to the length of the wave. If the slit is large compared to the length of the wave, the wave passes through the slit with slight curving at the edges. In fact if the wavelength is very short compared to the width of the slit, the wave goes practically straight through the slit with very little curving.

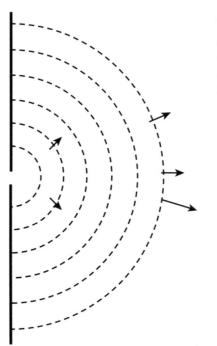

Figure 21.7

Straight waves passing through a slit bend sharply if the slit is small compared to the length of the wave.

The light passing through the slit with very little curving is similar to light passing by an object and casting a sharp shadow. That is a real possibility but the behavior of very short wavelengths causing a sharp shadow suggests something else. If light is a wave, then the wavelength of light must be very short because of the sharp shadows cast by even small objects. If you think about that, light passes through our eyes through an opening of maybe half a centimeter. We see clear images of light bouncing off objects and entering our eyes. That means that if light is a wave it must have wavelengths of much less than half a centimeter. Otherwise, we would have a large amount of curving and hence blurry images.

Those of us with healthy eyes see nice clear images from light passing through the pupil of the eye. The pupil is even less than a centimeter in diameter most of the time. If light is a wave, the wavelength of light must be very small compared to the diameter of the pupil of the eye. Diffraction is the bending of waves, the amount of bending depending on the opening through which it passes. If the opening is small, the slit or opening can act like a point source of waves. Sometimes when I can't find my glasses right away, I will make a very small hole by curling my finger into an

A Johnnie's Alert

Diffraction is an important property that waves share with light. Among other things, it provides the key to checking to see if you can measure the wavelength of light if light is a wave.

"O" shape. If I look through the small hole, I can see letters as if they were clearer. This is an example of light curving through small openings to make the letters appear clearer for me. You can see that this physics can come in handy sometimes.

Interference

The wave model explains most of the things that the particle model does, and explains refraction and diffraction even better. It appears to be a good model for light. Does it predict anything that might be checked like the particle model did in predicting light pressure? When you considered the standing wave pattern, you were asked to use your imagination to try to visualize a similar situation in two dimensions. Do waves in two dimensions exhibit the same behavior?

The standing wave was generated by a source and its image through a junction. The image had to be creating waves with the same frequency and at the same time. Two sources of periodic motion that are creating waves at the same time are said to be in *phase*.

Plain English

The **phase** of an object is the position and direction of movement of the object undergoing periodic motion. Waves **interfere** with each other when they move through each other and either reinforce each other or destroy each other. Waves **interfere constructively** when they interact to build each other up or reinforce each other. **Destructive interference** occurs when waves interact to cancel each other out or destroy each other. An interference pattern is possible only if you have two point sources in phase with the same frequency.

If you want to generate waves that behave like the waves causing the standing wave, you need two sources of waves that are in phase and have the same frequency. The reason for that is the standing wave had regions where crest met crest and trough met trough creating a deeper trough and a higher crest that are called loops. Waves interacting in that way are said to *interfere constructively*. The nodes in the standing wave are regions of *destructive interference*.

Dipping the finger into the water suggested that a point source of waves would dip into the water to cause circular waves. The ripple tank is equipped with two plastic beads mounted on a rocking arm that causes the beads to dip into the water in phase. The rocking arm is driven by a small electric motor that has a frequency that is the frequency of the dipping plastic beads. The frequency of the motor is measured with a calibrated strobe light. While the frequency is being measured, the wavelength of the water waves generated by the dipping beads is determined at the same time.

The beads dip into the water creating circular waves from two point sources. You know that the waves travel at the same constant speed in a single medium. The waves from the two point sources travel through each other and they create regions of higher waves and regions of perfectly still water. The total picture is called an interference pattern that looks very much like the two-dimensional standing wave pattern I asked you to imagine earlier.

Lines of still water and lines of maximum disturbance, deeper troughs and higher crests, characterize the interference pattern. The lines of still water are called *nodal lines*. The nodal lines are the regions where a crest from one point source meets a trough from the other.

Regions of maximum disturbance occur where a crest from one source meets a crest from the other source and a trough meets a trough. The drawing in Figure 21.8 helps you to visualize the interference pattern of two point sources, in phase and with the same frequency, in two dimensions.

Newton's Figs

Viewing light reflected from the surface of a ripple tank when it reveals an interference pattern shows regions where the water is perfectly still. Regions of water that have maximum disturbance separate those regions. The regions of still water are the nodal lines, destructive interference, and the regions of maximum disturbance are regions of constructive interference.

Plain English

Nodal lines are the regions in an interference pattern where the waves from two point sources, in phase and with the same frequency, cancel each other out. The destructive interference takes place where a trough from one source meets a crest from the other source.

Figure 21.8

The nodal lines separated by regions of maximum disturbance of the medium characterize the two-dimensional interference pattern.

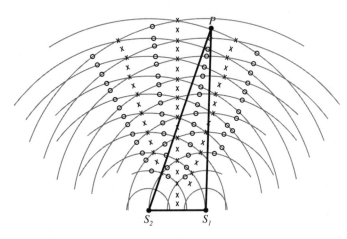

A Geometric Model of an Interference Pattern

I used a drawing compass to draw concentric circles around two point sources, S_1 *and* S_2, to represent circular waves generated by two point sources. The wavelength for each source is one centimeter and the distance between sources is three centimeters.

Newton's Figs

The perpendicular bisector of the line segment joining the point sources is in the middle of a region of maximum disturbance.

It is a most instructive activity if you make this construction for yourself drawing as many concentric circles as you can on a regular sheet of paper. The dark lines represent crests, and halfway between two crests from the same source is a trough. I marked where a crest from one source meets a trough from the other with O and where crest meets crest and trough meets trough with X.

Notice that there is a region of maximum disturbance of the water along the perpendicular bisector of the line segment joining S_1 *and* S_2. The nodal lines and lines of maximum disturbance are symmetric about the perpendicular bisector of the line segment connecting S_1 *and* S_2. You can then start at the perpendicular bisector and number the nodal lines 1, 2, ... If you make your own drawing and identify as many nodal lines as you can, you find a special relationship between a point on a nodal line very far away from the sources and S_1 *and* S_2.

I drew PS_2 *and* PS_1 for you and measured them to show you that relationship. PS_1 is 9.5 cm long and PS_2 is 10 cm long and the difference in the two is 0.5 cm or $\dfrac{\lambda}{2}$.

P in Figure 21.8 is on the first nodal line. You find that the difference in path length, PS_2-PS_1, for nodal line number two is $\dfrac{3\lambda}{2}$. In general, the difference in path length is $\left(n-\dfrac{1}{2}\right)\lambda$. The reason that is true is that a crest from one point source meets a trough from the other at a point of destructive interference. The only way that can happen is if the difference in path length, PS_2-PS_1, is an odd integral multiple of half wavelengths. If this reasoning is correct, there should be a way to measure the wavelength from the interference pattern and compare that with the direct measurement with the strobe light. The reason that measurement is important is that if light is a wave, the wavelength of light can be determined from an interference pattern.

Calculating the Wavelength Using an Interference Pattern

A blow-up of the diagram in the last figure is presented in Figure 21.9. Geometric quantities are identified and quantities that can be measured are labeled.

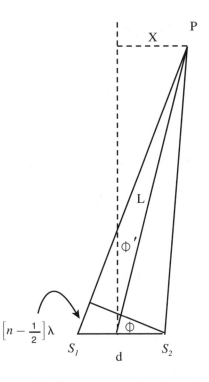

Figure 21.9

The geometry of the location of a nodal point enables you to calculate the wavelength of the waves.

The perpendicular bisector of the line segment joining S_1 and S_2 is shown as a vertical dotted line. The distance between S_1 and S_2 is labeled d and the distance from the foot of the perpendicular bisector of d to P is labeled L. The distance from P to the perpendicular bisector of d is labeled x. Two acute angles of the two right triangles are labeled ϕ and ϕ'.

If P is very far away from S_1 and S_2 compared to the distance between the sources, the two line segments drawn to P from S_1 and S_2 as well as the one from the foot of the perpendicular bisector of the line segment joining S_1 and S_2 to P, are parallel or nearly parallel.

That means that a perpendicular to PS_1 from S_2 cuts off a line segment that is the difference in path length, $PS_2 - PS_1$, and is labeled $\left(n - \frac{1}{2}\right)\lambda$. That approximation is meaningless unless P is very far away from S_1 and S_2. The acute angles are equal because two angles are equal if they have their sides perpendicular right side to right side and left side to left side. Since the angles are equal, their sines are equal so:

CAUTION

A Johnnie's Alert

The approximation that the three lines are parallel is good only if P is very far away from the point sources compared to the distance between them. Very far away would be at least ten times the distance between the sources of waves.

$$\sin \phi = \sin \phi'$$

$$\frac{\left(n - \dfrac{1}{2}\right)\lambda}{d} = \frac{x}{L} \quad \text{Definition of the sine of an angle.}$$

$$\lambda = \frac{dx}{L\left(n - \dfrac{1}{2}\right)}$$

The interference pattern enables you to measure x and L, and you can determine n by counting on one side of the perpendicular bisector of d. The value for the wavelength of the wave determined from the interference pattern is the same as the value found by measuring the length of the wave directly with the strobe light. If light is a wave, the wavelength may be determined in much the same way using an interference pattern.

An Interference Pattern of Light

Young gave us the tool needed to view an interference pattern for light as well as the interference pattern in the ripple tank.

Remember that two point sources in phase with the same frequency are needed to set up an interference pattern. In Young's experiment for light, one light source is used but it is viewed through two tiny slits very close together. The two slits act like two point sources just as you found in the diffraction discussion. The two point sources are in phase and have the same frequency because the same wave from a source passes through both slits at the same instant.

CAUTION

A Johnnie's Alert

Young's experiment gave you the idea of two tiny slits a small distance apart diffracting a sample of the same wave from the same source to create an interference pattern. The slits act like two point sources in phase with the same frequency.

You must use your imagination again to know what to look for if light is a wave. Suppose you take the pattern in Figure 21.8 and rotate the pattern about the line segment joining S_1 and S_2. Can you imagine nodal planes being generated by the nodal lines? If you view the nodal planes of an interference pattern of light, the nodal planes intersect the retina of your eye in lines and you see the lines of intersection. The lines are black because they are nodal planes that are regions of destructive interference. Bright regions, because of constructive interference of light, separate the black lines.

The diagram in Figure 21.10 helps to visualize the observed interference pattern and the experimental arrangement for viewing the nodal lines.

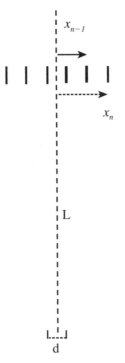

Figure 21.10

The nodal lines in the inter-ference pattern for light are seen as black vertical lines separated by bright regions.

When two point sources generate an interference pattern in the ripple tank, it is found that the closer together the point sources are the fewer the number of nodal lines. That means that the point sources of light must be close together so that we can distinguish the nodal lines. It also means that the slits that serve as point sources are very narrow because you want as much diffraction by each slit as you can get. My students painted one side of a microscope slide with a quick-drying graphite suspension. When the graphite was dry, they held two double-edged razor blades together and scraped two slits in the graphite with the edges of the blades. A thin line of graphite separated the slits by a distance d.

That distance was one half of the width of the two razor blades held together. The distance d is measured with a micrometer caliper accurate to 0.01 mm. The source of light is placed about 2.0 m away against a wall. The light source is viewed through the tiny slits by holding the microscope slide close to the eye. The interference pattern in Figure 21.10 is observed having up to 10 nodal lines.

The student observing the nodal lines can see the hand of a second student that marks positions on the wall. The student observer guides the student to mark a nodal line, then counts as many nodal lines as he can see clearly and instructs the recorder to mark the second observation.

A Johnnie's Alert

The interference pattern of light that you see is actually formed on the retina of your eye. The measurements you make in the laboratory are in a triangle similar to the triangle in your eye. That means that the results you obtain are reasonable.

The distance between slits is so small that the exact arrangement as used in Figure 21.9 is impossible to use. The same idea can be used to calculate the wavelength if light is a wave. You probably noticed that the nodal lines are the same distance apart. Even though you cannot measure x as you did in Figure 21.9, you can use that same information to calculate Δx, the distance between adjacent nodal lines. Let x_n represent the distance to the nth nodal line and x_{n-1} represent the distance to the nodal line next to it. Then

$$\Delta x = x_n - x_{n-1}$$

$$\Delta x = \frac{L\lambda\left(n-\frac{1}{2}\right)}{d} - \frac{L\lambda\left((n-1)-\frac{1}{2}\right)}{d}$$ By substitution because the results of the analysis of Figure 21.9 shows that $x = \dfrac{L\lambda\left(n-\frac{1}{2}\right)}{d}$.

$$\Delta x = \frac{L\lambda}{d}\left(\left(n-\frac{1}{2}\right)-\left((n-1)-\frac{1}{2}\right)\right)$$ By factoring out the common factor of $\dfrac{L\lambda}{d}$.

$$\Delta x = \frac{L\lambda}{d}\left(n-\frac{1}{2}-n+1+\frac{1}{2}\right)=\frac{L\lambda}{d}$$ By simplification.

The expression from the analysis of Figure 21.9 simplifies to $\Delta x = \dfrac{L\lambda}{d}$ using the arrangement in Figure 21.10. The distance L is not measured the same way as in Figure 21.9 because that distance is very far away from the microscope slide when it is compared to d, the distance between slits. The perpendicular distance from the slide to the wall that appears to contain the interference pattern is used for L. With proper supervision, you can shine a laser beam through the slits and view the interference pattern on a screen or the wall. The measurement of larger distances Δx and L will give you even better results. That is assuming you have access to a laser and proper supervision of its use.

The experiment reveals that blue light has an interference pattern with nodal lines measurably closer than the nodal lines in red light. That is, the experiment suggests that white light is made up of different colors of light. The red light is found to have a wavelength of about 600 nanometers and the wavelength of blue light is about 400 nanometers. The wave model for light explains everything that the particle model does at this point, and it does a better job explaining refraction and diffraction than the particle model does.

Notice that the work is not complete because you have not checked to see if the wave model explains light pressure. You can do that in time, but do not worry about that because you have two models to use to explain light. Use the wave model of light to explain those things that you can explain with it. Use the particle model to explain things that you can explain with it. The important thing is that you recognize that these are models to help you gain more understanding about the physical world. Whenever you hear that someone suggests that light is a particle or that light is a wave, you have a lot of knowledge that pops into your mind about the model they use to explain an idea.

> **A Johnnie's Alert** _____
> Go through each step of the algebra involved in the calculation of Δx until you can reproduce it on your own. It is not complicated but my students tell me that the details slip by them the first time through.

The Least You Need to Know

♦ Identify the different types of reflection of a wave in a spring.

♦ Explain how a wave in the ripple tank refracts when it crosses a junction from deep water to shallow water.

♦ Name the conditions that must exist for two point sources to generate an interference pattern.

♦ Describe the calculation of the wavelength of a water wave from an interference pattern in the ripple tank.

♦ Discuss the use of an interference pattern that enables you to calculate the wavelength of light.

Appendix A

Glossary

acceleration   The rate of change of velocity or the change in velocity divided by the time for that change to take place.

acoustics   The sound-producing qualities of a room or auditorium.

actual mechanical advantage   The gain a user realizes from a machine because it includes the effects of friction. (A ratio between forces.) Ideal mechanical advantage is the theoretical gain a user expects from a machine because it does not include the effects of friction. (A ratio between distances.)

adhesive forces   Forces of attraction between particles of different kinds of matter.

alpha particles   (^4_2He) Fast-moving Helium nuclei that have a positive charge and are emitted from the nucleus of radioactive elements like radium.

alternating current   (AC) An electric current that travels in one direction and then the opposite direction with a fixed period. Your local power company, through outlets in your home, provides AC.

ampere   The practical unit of electric current. If one coulomb of charge passes through a cross section of a conductor at a point in an electric circuit in one second, then the current is rated at one ampere.

amplitude   The maximum displacement (from its equilibrium position) of the object undergoing simple harmonic motion.

anode   The positive electrode of a cell or battery.

atom   The smallest particle of element that exists alone or in combinations.

atomic mass number   The number of nucleons the atom contains.

atomic number   The mass of an atom relative to that of the isotope carbon twelve. The whole number nearest the atomic mass is the number of nucleons the atom contains.

average speed   The magnitude of the change in position of an object that does not change direction, divided by the change in time required to make that change in position; or the distance an object travels divided by the time required to complete the trip.

battery   A combination of cells. It is constructed to overcome the limitations of one cell. A battery provides a larger current, a larger potential difference, or both.

beta particles   High-speed electrons emitted from the nucleus of radioactive elements. They have a negative charge.

branch (of a circuit)   A division of a parallel part of the circuit.

British thermal unit   The quantity of thermal energy required to raise the temperature of one pound of water one Fahrenheit degree.

calorie   The quantity of thermal energy needed to raise the temperature of one gram of water one Celsius degree.

candela   The unit of luminous intensity.

capacitor   An electrical device used to store an electrical charge.

cathode   The negative electrode of a cell or battery.

center of curvature   The center of the sphere of which the mirror is a part.

centripetal acceleration   The center-seeking acceleration resulting from uniform circular motion. Furthermore, it is the only acceleration defined by uniform circular motion.

centripetal force   The center-seeking force on an object at every instant that deflects the object into a circular path at a constant speed. At all times it is perpendicular to the direction of motion.

closed polygon method   A way of adding vectors that allows any number of vectors to be added together by joining them together foot to head until the sum is complete. The resultant vector is found by closing the polygon with a vector drawn from the foot of the first vector in the sum to the head of the last, in that order.

cloud chamber   A transparent container filled with alcohol vapor suspended between electrically charged plates. While a radioactive source emits alpha and beta particles, a source of light is directed at the alcohol-filled chamber. Trails much like vapor trails of jet planes are formed along the path where the radioactive particles have traveled. Their trails are easily viewed by the light reflected from the droplets of vapor in the cloud formed.

cohesive forces   The attractive forces between particles of the same kind.

components (of a vector)   Those parts whose sum is the given vector.

compression   That part of a longitudinal wave where the particles of the medium are pushed closer together.

concave spherical mirror   A segment of a sphere with the inside surface polished to reflect light.

condensation   The process of a vapor or a gas changing to a liquid.

conduction   The transfer of thermal energy within a substance from one particle to the next when the particles are not moving from one place to another. Thermal energy causes the particles to vibrate with greater energy, and the vibrating particles bump into neighboring particles, transferring the energy throughout the object.

conductors   Materials that allow electrons to move freely throughout the material.

convection   The transfer of heat by the movement of matter.

convex spherical mirror   A segment of a sphere with the outside surface polished to reflect light.

corona discharge (brush discharge)   A bluish glow of ionized gases formed at any sharp point of a conductor that is under the influence of high potential energy.

coulomb   The unit of electrical charge. It is equal to the charge on 6.25×10^{18} electrons.

covalent bonding   The combination of two atoms to form a molecule by sharing a pair of electrons.

crest   That portion of the graph of a transverse wave that lies above the time axis.

critical angle   The angle of incidence in the substance that has an angle of refraction of 90° in a second substance. (Only occurs with light as the light leaves a medium where it has a slower speed and enters a medium where it has a faster speed.)

cycle   One complete trip for an object moving in a circular path at a constant speed as well as the corresponding trip of its projection on the diameter of the circular path.

defining equation   A statement of a relationship between two units of measurement.

density   The amount of matter in a unit volume. Since matter is measured in two different ways, there are two types of density. Mass density is the amount of mass in a unit volume of matter and weight density is the amount of weight in a unit volume of matter.

derived quantities   Physical quantities that are defined using two or more fundamental quantities or one fundamental quantity used more than once.

destructive interference   Waves interact to cancel each other or destroy each other.

deuteron   The nucleus of an isotope of hydrogen made up of one neutron and one proton.

diffraction   The curving of straight waves as they pass through an opening or pass by a sharp corner.

diffusion   The movement of particles of one kind of matter into the empty space of a different kind of matter because of the random motion of the particles.

direct current (DC)   An electric current that travels in only one direction in the circuit. It travels from the negative side of the cell, through the circuit external to the cell, to the positive side of the cell. Conventional current is in the opposite direction.

displacement   A vector defined as a change in position.

Doppler effect   The apparent shift in the pitch of a source of sound because of the relative motion between the source and the observer.

dyne   The force required to accelerate a mass of one gram at a rate of one centimeter per second squared.

efficiency   A fraction greater than zero and less than one that expresses what part of input work the output work amounts to. The efficiency is usually stated as a percent obtained by multiplying the fraction by 100%.

electric circuit   The conducting path for the flow of charge.

electric current   The flow of charge through a cross section of a conductor past a point in an electrical circuit for each unit of time. (The rate of flow of charge.)

electron   A particle that has a charge of 1.6×10^{-19} *coulomb* and a mass of 9.1×10^{-31} *kg*.

element   A type of matter like copper, hydrogen, or neon that cannot, by chemical means, be broken down into simpler forms of matter.

equal vectors   Vectors that have equal magnitudes and the same direction.

equilibrium   When the sum of the forces is zero. If one force is used to balance the effects of two or more other forces, then that force is called the equilibrant and is equal in magnitude and opposite in direction to the resultant of the other forces.

evaporation   The process of a liquid changing to a gas or a vapor. It can happen at many different temperatures.

focal point (of a concave spherical mirror)   A point on the principal axis where all rays intersect after they are reflected from the mirror. That is true if they are rays that are parallel to the principal axis and near the principal axis. The focal point is halfway between V and C. That means the distance from V to C is $2f$ where f is the focal length of the mirror, the distance from V to F.

focal point (of a double convex lens)   A point on the principal axis of the lens where all rays intersect after being refracted by the lens. That is true if they are rays that are parallel to the principal axis and near the principal axis. Since refraction is reversible, and works in either direction through the lens, the lens has a focal point on each side that is the same distance from the center of the lens.

force of gravity   The downward pull on any mass placed above or on the earth's surface. This is a pull at a distance because nothing is attached to the mass pulling it toward the earth.

freely falling body   An object that is influenced by the force of gravity alone.

freezing point   The temperature at which a liquid changes to a solid at standard pressure.

frequency   The reciprocal of the period; given by $f = \dfrac{1}{T}$ and has units of $\dfrac{cycles}{s}$ or just $\dfrac{1}{s} = s^{-1}$ since a cycle is not a unit of measurement. Another common unit of measurement of frequency is the Hertz, where $1\,Hz = 1s^{-1}$.

fundamental quantities   The building blocks for the foundation of physics. Time, space, and matter are the quantities required to study that area of physics called Mechanics.

fusion   A name given to the process of changing from a solid to a liquid. It is more commonly called melting. Fusing means to bring together; when metals were melted, they were brought together to make alloys.

galvanometer   A meter connected in series in an electrical circuit to indicate the presence of a current. It also indicates the direction of the current in the circuit.

gamma rays   Very penetrating short-wavelength electromagnetic radiation emitted from the nucleus of radioactive elements.

graphic solution   Achieved by using drawing instruments to construct a scale drawing. A ruler provides the measurement of magnitude to scale and the protractor enables you to measure angles.

gravitational field   That region of space where the force of gravity acts on a unit mass at all locations on or above the surface of the earth.

half-life   The time required for half of the atoms in a given sample of a radioactive element to decay.

heat   The internal kinetic energy and potential energies of the particles of matter.

heat capacity   The quantity of thermal energy required to raise the temperature of a body one degree.

heat of fusion   The quantity of thermal energy required to change a unit mass or weight of a solid to a liquid at the normal freezing point without changing the temperature.

heat of vaporization   The quantity of thermal energy required to change a unit mass or weight of a liquid to a gas or vapor at the normal boiling point without changing the temperature.

hydrometer   An instrument used to measure the specific gravity of a liquid.

impulse   A physical quantity that results from a force being applied to a body for a certain amount of time which is equal in magnitude to the product of their values.

inertial frame of reference   A frame of reference that is at rest or moving at a constant speed. For our purposes, the earth is an inertial frame of reference.

instantaneous velocity   The velocity of an object at any instant of its motion. The instantaneous velocity vector is tangent to the path of motion of the object at every point along the path.

insulator   A material that does not allow freedom of movement of electrical charge.

intensity of illumination   The rate at which light energy falls on a unit area of surface.

interfere   Waves that move through each other and either reinforce each other or destroy each other. Waves interfere constructively when they interact to build each other up or reinforce each other.

ion   A charged particle caused by an atom gaining or losing electrons to exhibit an excess or a deficiency of electrons.

ionic bonding   The formation of a unit by the transfer of an electron from one atom to the other creating two ions more stable than the atoms were before the transfer. The two newly formed ions attract each other, due to opposite electrical charge, thus forming a stable unit.

junction   Where two media are joined together. That is, a spring can be anchored at a junction, two springs may be attached at a junction, or a spring may be attached to a string at a junction, or oil may float on water forming a junction.

kilocalorie   The quantity of thermal energy required to raise the temperature of one kilogram of water one Celsius degree.

kinetic energy   The energy an object has due to the fact that it is in motion.

law of heat exchange   The thermal energy gained by cold substances is equal to the heat lost by hot substances. (The First and Second Laws of Thermodynamics.)

longitudinal wave   A disturbance traveling through a medium in which the particles vibrate in paths parallel to the direction the wave is traveling.

loop (or anti-node)   The region of reinforced amplitude in a standing wave pattern.

luminous intensity   The strength of a source of light measured in a unit called the candela.

lumen   The unit of intensity of illumination.

magnet   An object made of iron or steel that is characterized by a north pole and a south pole each having the ability to strongly attract iron.

magnetic compass   A device used to find direction on the surface of the earth. The tip of the needle points to geographic north and the other end points to geographic south.

magnetic field   The region near a magnet containing all of the magnetic lines of force. The magnetic field is to a magnet what the gravitational field is to a planet.

magnetic force   A force resulting from the attraction or repulsion of a magnetic pole.

magnetic lines of force   Imaginary lines directed from the north pole through the space near a magnet into the south pole.

matter   Anything that occupies space and has mass.

maximum height (of a projectile)   The greatest distance the projectile reaches above the earth or the level of launch.

melting point   The temperature at which a solid changes to a liquid at standard pressure.

metallic bonding   The name of the attraction atoms of solid metals have for each other in a closely packed arrangement due to the continuous exchange of loosely held electrons in the outer energy levels of the atoms of the metal.

mole   Avogadro's number of items or a basic unit of quantity. 6.02×10^{23}.

molecule   The smallest particle of a compound.

momentum   A physical vector quantity that has its magnitude determined by the product of its mass and velocity.

move   An object experiences a change in position.

neutron   A particle with no charge that has about the same mass as the proton.

newton   The force required to accelerate a mass of one kilogram at a rate of one meter per second squared.

nodal lines   The regions in an interference pattern where the waves from two point sources that are in phase and with the same frequency cancel each other out. The destructive interference takes place where a trough from one source meets a crest from the other source.

node (or nodal point)   The points in a standing wave pattern where the superposition of the waves produces zero amplitude. Nodes are half a wavelength apart, as are the loops of a standing wave.

normal   A line perpendicular to a line or to a surface.

nucleus   The core of the atom that is made up of protons and neutrons.

parallelogram method   A way of adding vectors that requires the feet of two vectors to be located at the same point. A parallelogram is then constructed by drawing a line parallel to one of the vectors through the head of the other. A second line is constructed parallel to the second vector passing through the head of the first. The resultant vector is found by joining the feet with the opposite vertex of the parallelogram, in that order.

period   The time for the object in simple harmonic motion to complete one cycle.

phase   The position and direction of movement of the object undergoing periodic motion.

pith   A very light, dry, fibrous material. Pith is the central column of spongy cellular tissue in the stems and branches of some large plants. It is used with electroscopes to detect electric charge and is replaced by Styrofoam balls in some schools.

position   A vector used to locate a point from a reference point or frame of reference.

potential difference (between two points in an electric field)   The work done per unit charge as the charge moves between those points.

potential energy   The energy an object has because of its placement in a force field.

pound   The force required to accelerate a mass of one slug one foot per second squared.

pressure   A quantity determined by the force on a unit of area.

principal axis (of a concave spherical mirror)   The line that passes through the center of curvature and intersects the mirror at a point called the vertex of the mirror.

principal axis (of a lens)   The line passing through the principal focus and the center of the lens.

proof-plane   A small metal disk attached to an insulator. The disk may be rubbed on the charged insulator to sample the charge and limit the number of charges taken. The physical size of the disk limits the size of the charge to be placed on the electroscope.

propagation   The sending out or spreading out of a wave from a source.

proportion   An equation each of whose members is a ratio.

proton   A positively charged particle having a charge of the same magnitude as that of an electron with a mass of about $1.7 \times 10{-}27 kg$.

pulse   A wave or a disturbance of short duration.

quantitative description   Involves actual measurements such quantities as temperature, time, and current.

radian   The central angle of a circle subtended by an arc that is equal in length to the radius of the circle.

radiation   The transfer of thermal energy by having only the energy transferred. No substance or convection currents are needed.

range (of a projectile)   The maximum distance traveled horizontally by the projectile. It is measured from the point of launch to the point of return to the same level.

rarefaction   That part of a longitudinal wave where the particles of the medium are being spread apart.

reflection   The change in direction of light when it strikes a boundary causing the light to bounce off the surface of the boundary so that the angle of reflection is equal to the angle of incidence.

refraction   The change in direction of light when it leaves one medium and enters a different medium.

regelation   The melting of ice under pressure and then freezing again after the pressure is released.

resistance (in an electric circuit)   The opposition to the flow of charge or current in the circuit.

resistor   A component in an electrical circuit used to establish the amount of current and/or potential difference at different places in the circuit.

resolution   The name of the process of identifying the parts of a vector.

resonance (in sound)   The increased amplitude of vibration of an object caused by a source of sound that has the same natural frequency.

resultant vector   The name assigned to the vector representing the sum of two or more vectors.

ripple tank   A laboratory instrument used to study waves in two dimensions. It is a shallow tank that holds water about one to two centimeters deep. It has a glass or clear plastic bottom about 24 inches square.

scalar   A quantity that has magnitude only. Any units of measurement are included when we refer to magnitude.

scalar multiplication   The product of a scalar and a vector that results in a vector with a magnitude determined by the scalar. The direction of the product is the same as the original vector if the scalar is positive and opposite the direction of the original vector if the scalar is negative.

schematic diagram   A symbolic representation of electric circuits. That is, symbols representing components, conductors, and instruments of measurement are used instead of pictures of those items.

significant figures   Those digits an experimenter records that he or she is sure of plus one very last digit that is doubtful.

simple harmonic motion   The to and fro motion caused by a restoring force that is directly proportional to the magnitude of the displacement and has a direction opposite the displacement. Examples include the motion exhibited by a vibrating string or a simple pendulum.

simple pendulum   A physical object made up of a mass suspended by a string, rope, or cable from a fixed support. It is called a simple pendulum because the string, rope, or cable has a negligible amount of mass compared to the mass of the object being supported. A physical pendulum is a physical object having the mass distributed along the full length of the pendulum and not having most of the mass concentrated in one place as in the simple pendulum.

solidification   The process of changing from a liquid to a solid. Commonly known as freezing.

specific gravity   The ratio of the weight density of the liquid to the weight density of water. It is also the ratio of the mass density of the liquid to the mass density of water.

specific heat   The ratio of its heat capacity to the mass or weight of a substance.

spring balance   A device containing a spring that stretches when pulled by a force. The amount of stretch is calibrated to correspond to force units, newtons, dynes, or lb.

standing wave   The superposition of two waves of the same frequency moving in different directions in the same medium. A standing wave can be established in a medium of proper length. Resonance of sound in a closed pipe occurs for lengths of a quarter wavelength or odd integral multiples of quarter wavelengths.

sublimation   The direct change of a solid to a vapor without going through the liquid state.

superpose   When one wave adds onto the other. During the instant of superposition, neither pulse is recognizable individually because they appear as one combination of pulses.

surface tension   A condition of the surface of a liquid resulting from the attractive forces of the molecules of the liquid that causes the surface to tend to contract.

temperature   The condition of a body to take on thermal energy or to give up thermal energy. Bodies with high temperature are in a condition to give up thermal energy to cooler bodies. Bodies with low temperature are in a condition to take in thermal energy from warmer bodies.

terminal velocity   The maximum velocity of a falling object in air. The air friction on a falling body will increase with velocity until the drag of air friction is equal in magnitude to the force of gravity on the body.

thermodynamics   The study of quantitative relationships between other forms of energy and thermal energy.

time of flight (of a projectile)   The total time it is in the air.

trajectory   The arced or curved path of a projectile.

transverse wave   A disturbance traveling through a medium in which particles of the medium vibrate in paths that are perpendicular to the direction of motion of the wave.

trigonometry   A branch of mathematics that deals with relationships of angles and corresponding sides of triangles.

triple beam balance   A device for measuring mass by comparing an unknown mass with a known mass by balancing two pans. The two pans are attached to a beam that is supported by a fulcrum in much the same way as a seesaw or teeter-totter.

trough   The portion of the graph of a transverse wave that lies below the time axis.

unified atomic mass unit   The unit of mass used to stipulate nuclear masses. The atomic mass unit is equal to $\frac{1}{12}$ of the mass of an atom of carbon twelve.

uniform motion   Motion characterized by a constant speed. Whenever a situation in physics states or implies uniform motion, then you will know that the speed is constant. If the speed is constant, then you know that the object has uniform motion.

unit analysis   The process of defining a disguise of unity (one) in terms of units of measurement that will enable you to change from one unit of measure to a larger or smaller unit of measure without changing the value of the measured quantity.

valence electron   An electron in an incomplete energy level that an atom can lose and become more stable or for a different atom to gain to become more stable resulting in a more stable bond. These loosely bound electrons are largely responsible for the chemical behavior of an element.

Van de Graaff generator   A motorized source of a high concentration of electrons with high potential energy.

vector   A quantity that has magnitude, direction, and obeys a law of combination. The magnitude includes the units of measurement.

velocity   The rate of change of displacement. Velocity is a vector quantity that has speed as its magnitude.

voltmeter   The instrument used to measure potential difference in an electrical circuit.

wavelength (of a transverse wave)   The distance from the beginning of a crest to the end of an adjacent trough. One wave is made up of a crest and a trough. The distance from the beginning of a crest to the end of the same crest is one half of a wavelength.

weight (with respect to the earth)   The force of gravity on any object. It is a vector quantity with direction radially inward toward the center of the earth. The direction as described locally is down.

Appendix B

Answers to Physics Phun

Chapter 1

Problem Set I

1. 2
2. 3
3. 1
4. 4
5. 1

Problem Set II

1. $2.5 \times 10^5 \, mi$
2. $4.83 \times 10^6 \, m$
3. $2.3 \times 10^7 \, s$
4. $8.76 \times 10^{-4} \, kg$
5. $1.0003 \times 10^0 \, cm$

Problem Set III

1. 951.2 cm
2. 324.0 ft
3. 76.7 cm

Chapter 2

1. $1.5 \times 10^{11} \, m$
2. $8.64 \times 10^4 \, s$ or $8.6400 \times 10^4 \, s$ because the digits in defining equations are all significant.
3. $x = v_0 t + \dfrac{1}{2} a t^2$

Chapter 3

1. 2.6 cm $E30°S$ or $S60°E$
2. The difference vector $\vec{B} - \vec{A}$ is the opposite of $\vec{A} - \vec{B}$.
   $\vec{B} + \left(-\vec{A}\right) = \vec{B} - \vec{A} =$
   $3.5S10°W$ or $3.5W80°S$

Chapter 4

1. 2.3 m/s
2. 5.5 s
3. 4.5 m

Chapter 5

Problem Set I

1. $3.0 \times 10^1 N$
2. $68 ft / s^2$
3. 16.4 N $S36°W$ or $W54°S$

Chapter 6

Problem Set I

1. 16 ft, 48 ft, 64 ft
2. 20.4 m, 2.04 s

Problem Set II

1. 160 lb
2. 5.0×10^1 kg

Chapter 7

Problem Set I

1. $y_1 = 57 ft, y_2 = 82 ft, y_3 = 75 ft,$
   $Y = 83 ft$
2. $Y = 39.4m, y_3 = 39.3m$

Problem Set II

1. $T = 5.10s, Y = 31.9m, X = 221m$
2. $t_{max\,ht} = 44.4s, \ a = 9.80\dfrac{m}{s^2},$
   $v = v_{0H} = 435\dfrac{m}{s}, F_{net} = 1.12N,$
   $t_{max\,ht} = 44.4s, \ T = 88.8s, Y = 9650m,$
   $X = 38,600m$

Chapter 8

Problem Set I

1. 1.0 s
2. $979 \dfrac{cm}{s^2}$ or 9.79 m/s²
3. 24,000 dynes or 0.24 N
4. $\dfrac{2}{3}\dfrac{1}{s}$
5. The acceleration is maximum when the displacement is maximum. The acceleration is zero when the velocity is maximum.

Problem Set II

1. $2.56 \times 10^{-7} N$
2. 3.4 N, the scale would have a higher reading at the North Pole because there is no centripetal force.

Problem Set III

1. $8.43\dfrac{m}{s^2}$
2. 443 N
3. $5.96 \times 10^{24} kg$

Chapter 9

Problem Set I

1. 334 ft-lb

2. 1870 ft-lb

3. 2780 joules

4. (a) 240 N; (b) No work is done because there is no displacement in the direction of the centripetal force.

5. 820 ft-lb

Problem Set II

1. (a) $W_{k\,out} = 1300\,joules$;
   (b) $W_{k\,in} = 1700\,joules$;
   (c) efficiency= 76%

2. a little less than 310 N

Chapter 10

Problem Set I

1. (a) 93 ft-lb

2. (a) 1.48×10^5 ft-lb; (b) 176 lb; (c) 23.4 hp assuming one horsepower is 550 ft-lbs/s

3. 4.56×10^{-10} ergs

Problem Set II

1. (a) 44.9 m/s; (b) 6880 joules; (c) You might expect the ball to make a dent in the earth.

2. (a) 242,000 ft-lb; (b) 242,000 ft-lb; (c) The brake did the work along with the tires sliding on the road.

Chapter 11

1. 52,100 ft-lb, 70.2 mi/hr or 103 ft/s

2. 7.67 m/s

Chapter 12

Problem Set I

1. 4360 lb

2. 15.0 N-s, 3.00 m/s

3. 8.90 N

Problem Set II

1. 0.247 m/s in the opposite direction of the object

2. 17,100 kg

3. 0.71 m/s in the direction opposite the bullet

Chapter 13

Problem Set I

1. 23.5 N

Problem Set II

1. 4.54×10^4 N/m²

2. $9.8 \times 10^4\,lb$

3. 2.56 lb/in²

Chapter 14

1. $2.78 \times 10^4\,\dfrac{lb}{ft^2}$; No; 193 lb

2. 7.5

3. 3.11 ft

Chapter 15

Problem Set I

1. $29.7°C$

2. $74.3°F$

3. $0.10 \dfrac{Btu}{lbF°}$ to $0.11 \dfrac{Btu}{lbF°}$

Problem Set II

1. $79°C$

2. 2400 Btu

Problem Set III

1. $2.1 \times 10^4 \, Btu$

2. 34,600 cal

3. $104°F$

Chapter 16

Problem Set I

1. 0.333 m, 0.0222 s

2. 60 Hz, 0.0167 s

3. 16.5 m, 0.0165 m

Problem Set II

1. 20.4 s

2. 4.48 ft

3. 0.9 s

Chapter 17

1. 108, 47 protons, 61 neutrons, 108 g, $1.79 \times 10^{-22} \, g$

2. 82, 207, 207 g, $3.44 \times 10^{-22} \, g$

3. 79 protons, 79 electrons, 118 neutrons

Index